The Best Kits Catalog

Also by Frank Coffee

THE COMPLETE KIT HOUSE CATALOG
THE SELF-SUFFICIENT HOUSE
EVERYTHING YOU NEED TO KNOW ABOUT CREATIVE HOME FINANCING

The Best Kits Catalog

Top-of-the-Line, Time-Tested Kits for Constructing Practically Anything

FRANK COFFEE

HarperPerennial
A Division of HarperCollins*Publishers*

HarperCollins books may be purchased for educational, business, or sales promotional use. For information, please write Special Markets Department, HarperCollins Publishers, Inc., 10 East 53rd Street, New York, NY 10022.

FIRST EDITION

Designed by Barbara Cohen Aronica

Library of Congress Cataloging-in-Publication Data
Coffee, Frank.
The best kits catalog / Frank Coffee.—1st ed.
p. cm.
Includes index.
ISBN 0-06-096905-9 (pbk.)
1. Handicraft. 2. Do-it-yourself-work. 3. Industrial arts. I. Title. II. Title: Kits for making practically anything.
TT157.C525 1993
680′.28—dc20 92-53384

---

93 94 95 96 97 RRD 10 9 8 7 6 5 4 3 2 1

Contents

Introduction

Welcome to the wonderful world of kits.

We should warn you up front that kits can be addictive. Start with a simple clock or inexpensive furniture project, and before you know it you'll be looking at plan books for the big stuff, like sailboats and kit houses. Merely thumbing through this book could start you down the road to becoming a kit enthusiast. But we would encourage you to pursue that course, for building with kits, along with giving great personal satisfaction, almost always can save you money—sometimes a lot of money. Some kits might even suggest the way to a profitable business, or a sideline run from the home.

The emphasis here is on the practical; pure hobby kits have been largely ignored. But just turn a few pages and you'll soon discover that we've included all the best of an extraordinary range of projects—whether it's a hooked rug or a rocking horse to be made for gift-giving, the addition of a sauna or a sunspace to an existing property, or—the ultimate project—the building and furnishing of the home of one's dreams. There even are kits, from guitars to harpsichords, for the musically inclined, as well as kits for those whose driving ambition is to own a "classic" car or to build and fly their own aircraft.

Kit selections have been based on superiority of design, ease of construction, exceptional value, or, in some instances, simply because of the uniqueness of the item. In making our selections, we've consulted with other kit enthusiasts, how-to writers and editors, building contractors, family and friends, and the ever-helpful personnel at a good number of home centers. We've also drawn on our own nearly forty years as a writer/editor on kits, how-to, and the home, and more than fifty years as a kit builder. If we've missed a kit that clearly belongs here, and you can make a strong enough case for it, we'll consider it for inclusion in any subsequent editions of this book.

Most of the kits covered can be ordered direct from the manufacturer or from a closer-to-home source to whom you may be referred. For your convenience, we've included an address and telephone number for every manufacturer or other supplier mentioned in these pages. Since prices are subject to change, it's a good idea to confirm the price before ordering or sending payment for any of the kits featured here. Most kit manufacturers will send you a free brochure for the asking, but if you want a more detailed catalog or product information package, there's usually a charge, which we've noted.

Our special thanks to the many manufacturers who supplied photographs and artwork so that we could bring to the reader this coverage of the best in kits. We are also indebted for their help in the preparation of this book to Lorma Barnett, Barbara and Roger Courneen, Robert Dodd, David Gustafson, John Hagerty, Bill Halbach, John Turner Hartman, Kenneth James, Fred L. Jones, Peter Karagianis, Edward and Margret Lettieri, William T. McKeown, Bonnie McLennaghan, Bill Miller, Ben Thomas, and Diane Walden.

Home Furnishings

FURNITURE

Virtually any item of furniture made largely of wood can be purchased in kit form today for assembly and finishing by the consumer. With a kit, usually purchased by mail order direct from the manufacturer, you not only save up to half or more of the cost of a comparable piece of assembled and finished furniture bought at retail, but you end up with an original. Unlike furniture from a factory assembly line, no two pieces of hand-assembled, hand-finished furniture, even though cut to the same pattern, will ever be identical. The work that you do makes for a one-of-a-kind piece and will reflect, possibly for generations to come, your appreciation of handcrafting and the finer things in life.

Furniture in kit form is nearly as old as America itself. In the early years, peddlers sold furniture to homesteaders in pieces because it was more convenient to transport it that way. And the Shakers, over a century ago, were making furniture in parts in one community for shipment and subsequent assembly in other communities. But until a decade or so ago, kit furniture was pretty much limited to softwood pieces of fairly simple design. The recent trend in factory-direct kits, however, has been toward good-quality reproductions or adaptations of some of the finest examples of the cabinetmaker's art, at "reasonable" prices, in hardwoods. If you love antiques, reproductions copied line for line from classic colonial, Queen Anne, and Shaker pieces in museums and private collections are now included in the catalogs of manufacturers of fine furniture kits, and their assembly and finishing can be a lot more satisfying than being an underbidder on pieces that only the very rich can afford.

The big advantage with a furniture kit, other than the money saved, is that it allows the consumer to handcraft furniture pieces without having a workshop and a collection of expensive power tools. Most kits come with all wood parts precut, ready to assemble and finish. Where rabbets, miters, dovetails, or mortise-and-tenon joints are called for,

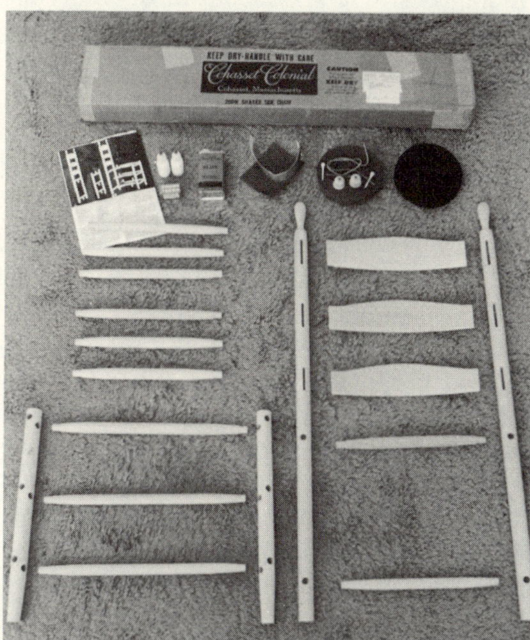

Before you start putting a kit together, lay out all the pieces, identify them, and make certain you have everything you need to complete the project.

they will have been machined at the factory. Door frames invariably come preassembled, ready for hardware and glass. With most kits, no special tools are needed. Nor is a conventional workshop area needed. A hammer and a screwdriver, and the family-room floor or a worktable, usually will do.

Even modestly skilled do-it-yourselfers can produce a handsome piece of furniture from most kits. Kits for the absolute novice generally are assembled with glue and a few screws. The most challenging projects might require the use of an electric drill, and such additional tools as a try square and woodworking clamps, though the latter probably could be improvised if you don't anticipate making enough use of them to justify their purchase.

All wood parts usually come from the factory presanded on all surfaces, but touch-up sanding is strongly recommended. Your finishing results can be no better than the effort you put into the final sanding of the wood parts. For a professional finish, you'll want to remove any minor nicks and marks left by the shaping machines. To avoid scratches, sand only in the direction of the wood grain. For a satin finish on exposed surfaces, use progressively finer grades of sandpaper, on the order of grit 220 and 240 garnet finishing paper and soft-backed fine (320) or very fine (400) silicon-carbide paper. Silicon-carbide paper will not scratch, yet cuts rapidly. Use a sanding block on the larger flat surfaces. Since changes in humidity can affect the fit of joints that passed inspection in the factory, some sanding may also be necessary with tenons (male), which should be neither too tight nor too loose in the mortise (female) before glue is applied.

Kits that come with screw holes predrilled further reduce the chance for error, and factory predrilling is especially welcome with hardwoods. It takes both strength and experience to drive screws in hardwoods without full-depth pilot holes. Even with pilot holes, apply a dab of wax or soap to the tip of each screw before inserting it. The lubrication can make the work of driving the screw a lot easier and reduces the possibility of splitting the wood. With softwoods, take care not to "overdrive" screws, or you may strip out the hole.

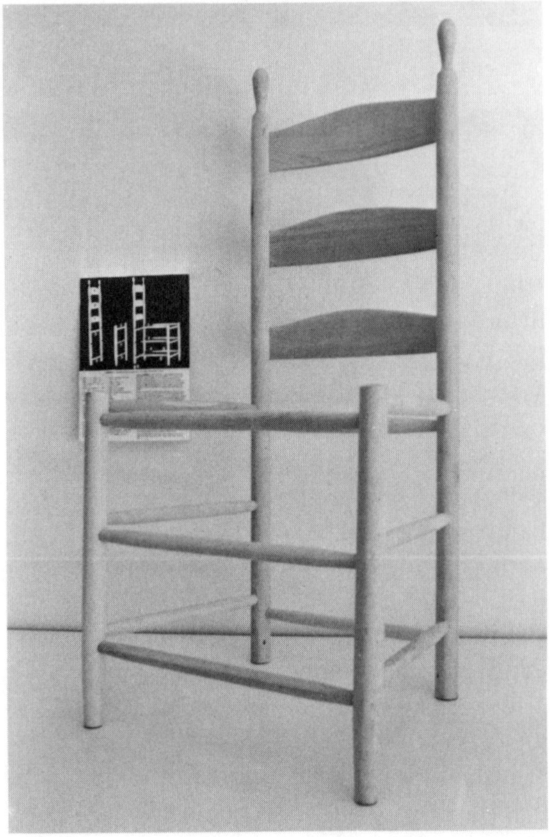

Go through a dry-run assembly, without glue or screws, sanding lightly where joints require it for fit. Joints should not have to be forced together.

Read through all instructions at least twice and study any drawings carefully before you start putting things together. Lay out all the pieces, identify them, and make sure you have everything you need to complete the project. Don't be in a hurry. With the glues that are used today, it's difficult to go back a step or two if you make an error in the assembly. Go through a dry-run assembly, without glue or screws, sanding lightly where joints require it for fit. Joints should not have to be forced together. If you have a lot of parts, it may be helpful to break the project down into sections or subassemblies, or number the parts for easy identification.

When it comes to the actual assembly, be careful when applying glue to the joints. Excess glue, even when wiped off with a warm, damp cloth, the usual recommendation, sometimes leaves a smear that is not easily detectable until stain* is applied. The glue forms a sealer that stain can't penetrate, and it stands out like a birthmark. Sanding is required to eliminate such smears. If you don't catch them before staining the wood, sand between coats. Better yet, allow any glue "overflow" from joints to dry, then remove it with a knife or chisel, and resand the area. This avoids pushing glue into the grain, which can happen when it's wiped while wet.

When using paste wood compounds to fill nicks and cracks, select a filler material colored to match the stain and apply it where needed *after* the first coat of stain. Filler material too generously applied also can lead to problems. Some fillers harden fast and are not easily sanded down. Be stingy with the use of filler; don't smear it on. Touch up as required with stain.

The instructions may call for the use of adjustable hand screws, pipe clamps, or band or web clamps to secure glued joints and subassemblies until the glue has set. If you don't have any conventional clamps, joints sometimes can be nailed (using brads) to hold them tight until the glue binds. Any brads used should be set with a nail punch, and the nailheads should later be covered over with filler material. Clamp nails, driven into saw kerfs to draw adjoining components together, can be used to secure miter joints with or without glue. Another alternative to store-bought clamps is to run a loop of rope around the assembly, twist it tight with a dowel, and then tie down the dowel.

Before staining, corners and any sharp edges should be rounded off or "broken" with a couple of strokes of fine sandpaper. Remove all sanding dust with a vacuum, dry brush, or lint-free cloth soaked with denatured alcohol. Some instructions recommend using a tack cloth for this purpose. But be careful here. If you use a tack cloth before staining, you could leave varnish smudges that will seal the wood, leading to a splotchy finish. A tack cloth does a good job of picking up fine dust particles, but if you intend to stain, restrict its use to the post-staining stages.

Tack cloths can be purchased at paint stores. They can also be homemade. Take a square of finely woven lint-free cloth, such as a man's handkerchief, dip it in tepid water, and wring it out. Then sprinkle the cloth lightly with varnish and wring it out again. It should be just tacky enough to pick up fine dust particles without depositing any grain-raising moisture. If the tack cloth is stored in an airtight container, it can be used again and again.

Pay close attention to the instructions printed on all finishing materials. With most stains,

*Stain brings out the grain and character of woods, even softwoods, and is incorporated in the finish of virtually all quality furniture, period reproductions especially. As with any fine piece of furniture, wood grain makes each piece unique.

for example, it will be pointed out that woods of great porosity, such as fir and ponderosa or sugar pine, should be primed before stain is applied, for a more uniform base. Always test the product on a hidden surface before you proceed with the finishing of the project. If you're not pleased with the result, you can always try a different shade, or another manufacturer's product. With the right choice of stain and proper application, you can make even inexpensive softwoods like pine look like the more expensive hardwoods.

It takes experience to achieve professional results without leaving brushmarks with brush-on finishes. If you haven't done any furniture finishing, we therefore recommend that you select stains, varnishes, and other finish coatings that can be applied with a brush or a soft rag and then be "ragged off" for easily controlled depth of color and no brushmarks. The newer finishing materials are much improved over products you may have used only a few years ago. Choose from wiping stains, clear urethane varnishes, or the new varnish stains, which let you stain and top-coat in a single step. Just be sure that the finishes are compatible with the wood to which they will be applied. Rub the finished surfaces lightly with 4/0 steel wool after each coat of stain or varnish has thoroughly dried to smooth any rough spots caused by dust or raising of the grain. Between applications, be sure to wipe down thoroughly to get rid of any steel-wool particles.

With smaller projects, it may be useful to apply all but the final finish before assembly. If this procedure is used, avoid getting the finishing material on any surfaces to which glue will be applied, or weak glue joints will result. Whether the project is small or large, wood surfaces should be finished on both sides to prevent warping and shrinking. Thorough finishing is also a sign of top-quality furniture.

One of our favorite pieces of furniture is a butler tray table that we built from a kit more than a dozen years ago. The finishing kit included paste stain and paste varnish, and we could not have been more pleased with the results. With total control over the intensity of color and depth of luster, we achieved a beautiful hand-rubbed finish—two coats of stain, five coats of varnish, and two coats of clear "bowling alley" paste wax—that will be a source of pride for many years to come.

If you've been to Colonial Williamsburg and other restorations where assembled and finished handcrafted furniture reproductions are offered for sale, you may have been shocked at the out-of-reach prices, such as $2,500 for a c.1780 Philadelphia rococo side chair, or $5,500 for an early New England dressing table or lowboy. In kit form, from manufacturers who will be introduced to you in the following pages, you can purchase fine examples of Early American and other classics for a whole lot less. Kit price for an excellent c.1770 brandy stand reproduction might run to $175, for a bowback Windsor armchair, $275, and for an unassembled, ready-to-finish Queen Anne lowboy, $450.

In spite of the label, Queen Anne furniture, named for the English queen who reigned from 1702 to 1714, appeared in the American colonies before it appeared in England. Originated in Europe by the Dutch, who borrowed it from the Chinese, the style was brought to America by the Dutch and was popular here for more than a century.

The ease of assembling and finishing a piece of furniture that is just as elegant as any sold in fine furniture stores will likely come as a revelation. But much more is offered in kit form than just the enjoyment of the look and feel of antiques without paying today's prices for antiques. Kit furniture includes everything from simple barrister bookcases and decorator tables to rolltop desks and armoires.

With most mail-order sources of kit furniture, orders can be placed by telephone, often to a toll-free 800 number, and charged to MasterCard, Visa, or American Express credit cards. Some companies also have a showroom or retail location that you may visit.

For our selections of the best in kit furniture, we've gone to seven different sources, representative of different periods, styles, and price ranges.

The Bartley Collection

Much of the credit for the high quality in ready-to-assemble-and-finish antique furniture reproductions must go to The Bartley Collection, which features line-for-line reproductions of some of the great English and American masterpieces of eighteenth-century cabinetry. Many of the originals of the Bartley reproductions are in museum collections and would fetch hundreds of thousands of dollars were they to come up at auction today.

The precut parts and hardware for many of the items sold through the Bartley catalog—as well as the hand-finished furniture pieces that Bartley also offers (at two to three times the price of the kit)—must meet with the approval of museum reproduction committees for attention to authenticity, detail, and workmanship. Brasses, for example, in some cases precisely copied from the original, are all solid and hand-finished. Stain colors are all authentic eighteenth-century wood tones, the darkness of which is controlled by the number of applications wiped on.

The eighteenth-century New England pencil-post bed is an American classic. Re-created by The Bartley Collection, in single, double, queen, and king sizes, it is kit-priced from $619. Both the comb-back Windsor chair, copied from an original made and signed by Joseph Henzy, a chairmaker made famous for the Windsors he crafted for Independence Hall, and the Bedside Bachelor's Chest, an exact reproduction of a 1790 Chippendale original, are also offered as Bartley kits, priced at $377 and $695, respectively. (*The Bartley Collection*)

Bartley reproductions are not inexpensive, but you can easily put them together from a kit and finish them yourself—even if you have few tools and no woodworking experience. Most Bartley Collection kits are offered in a choice of mahogany or cherry.

The sudden flowering of many unsuspected manual talents can be attributed to the gift or purchase of a timeless eighteenth- or early-nineteenth-century reproduction in kit form from The Bartley Collection. If you are new to The Bartley Collection, you might want to

Tapered legs reflect the Plantation Huntboard's Hepplewhite heritage. The large drawers, left and right, are actually doors opening into cupboards. It's a handsome piece, whether it's painted or finished with a traditional varnish; the kit price is $695. (*The Bartley Collection*)

start with a quick-and-easy project like the Five Board Bench ($119), an excellent addition to the mud room—or anywhere you need a place to sit and change your footwear, shed muddy gardening gear, or gather the day's luggage on your way out the door.

The single most popular kit for new Bartley collectors is its Hadleigh Tea Chest (also $119). This charming Chippendale-style chest dates from the time when tea was so expensive it was kept under lock and key. It is both decorative and practical and makes a perfect gift. Velvet jewelry inserts are available at $30 a pair. Small keepsake chests like this one are very popular and, at last count, Bartley offers at least eight different such kits, priced from $39 to $339. The highest price is for an elegant jewelry case with three sliding trays that can be locked behind paneled doors.

Other get-acquainted projects might include Bartley's Brandy Stand ($175), a lovely accent piece reproduced from a superb 1770 original; a graceful, one-drawer Queen Anne Nightstand ($319), or, one of our favorites, a Butler Tray Table ($375). The Bartley version of this classic table has been "copied" by others but usually without the craftsmanship reflected in the "floating" top, which is made up of eleven tongue-and-groove pieces.

In addition to the few we've chosen to show here, The Bartley Collection offers kits for many other equally fine pieces based on originals by such superb eighteenth-century cabinetmakers as Thomas Chippendale and George Hepplewhite. The Shakers also are represented, as is an unknown Connecticut cabinetmaker who, in about 1770, built the original on which the Bartley signature piece, a Bonnet Top Highboy, is based. The kit for this magnificent piece, which is 7½ feet tall, is $2,500.

The Bonnet Top Highboy, however, is not Bartley's most expensive kit. That honor goes to the equally magnificent, and nearly as tall, Newport Chest-on-Chest; kit price, $3,500. This is a reproduction of a c.1760 companion piece to a Newport secretary that was sold at auction on June 3, 1989, at Christie's in New York City, for $12.1 million!

For a demonstration project, we selected a kit much less costly, but no less authentic or pleasing to the eye: a Chippendale End Table. The kit cost $319, and we chose mahogany for our wood, since it was the wood most

Bartley's splendid Chippendale End Table includes many of the best features of eighteenth-century design, but adapts perfectly to twentieth-century needs. A terrific piece for the living room, it works just as beautifully as a bedside table. The kit price is $319.

often used by Chippendale. There was a shipping charge, with delivery by United Parcel Service (UPS), of $16, and we ordered a Bartley finishing kit, which contains ½ pint of gel stain, ½ pint of clear gel varnish, crack-and-gap filler, and sandpaper, for $12.95.

The kit came so well packaged we suspect the packaging itself had been worked out with a computer. We read through the instructions and other literature that accompanied the kit and then proceeded to identify all the parts and check them off against the parts list. We then read through the instructions once more before beginning our dry-run assembly. It still took some juggling of parts to get the legs and rails in their correct positions. With all the mortises, none of the legs are interchangeable.

Once we were satisfied that we knew what we were doing, we began with the final drawer assembly. Except for some sanding of the dovetails for fit, this was pretty straightforward. The next step, following the instructions, was to sand and *finish* the tabletop and the shelf. The drying time recommended between applications of stain and then the clear varnish gave us plenty of time to complete the finish sanding (though not much was needed) on all the other table parts.

To assemble and glue the legs and rails together, the shelf must be in place. This is the

Before beginning our assembly of the Chippendale end table, we went through a dry run, without glue, for familiarization and to check the fit of each joint, sanding tenons and dovetails lightly where joints would otherwise have been too tight.

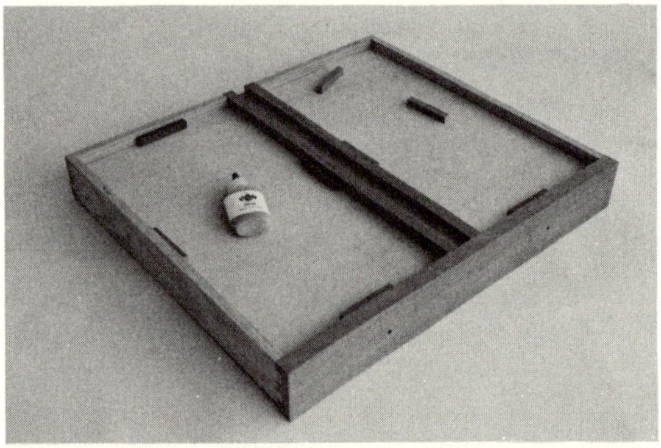

Assembly began by gluing one drawer side to the drawer front and back, and then sliding the drawer bottom into place before attaching the other side. The drawer guide track and glue blocks are added after allowing drying time for the glued-up drawer.

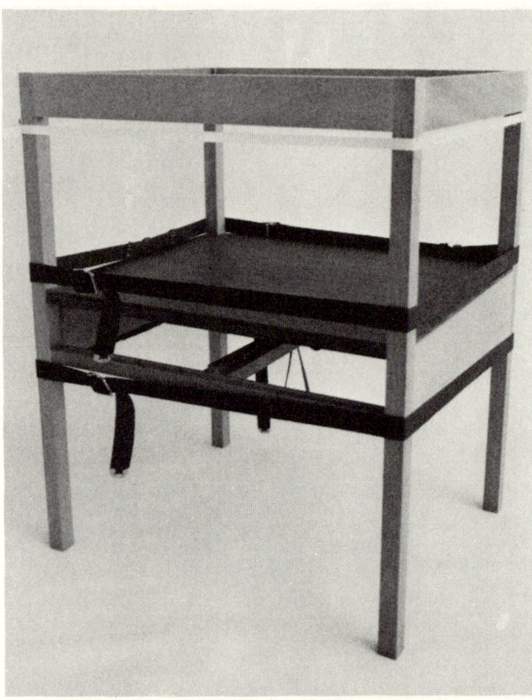

With the shelf in place, all four legs, all four top rails, and all four bottom rails were glued and joined in one operation. We then secured the assembly with band clamps, confirmed that the table was square, and allowed the glue to dry overnight.

After some final light sanding to remove a few glue spots, we applied two coats of Bartley's paste stain and then two of varnish to the table base. All that's needed now is to attach the finished top to the base with Z clips and insert the drawer.

only challenging stage of the assembly. All four legs, all four top rails, and all four bottom rails have to be glued and joined in one operation. We then secured everything with band clamps, confirmed that the table was square, and rested it on a flat surface to allow the glue to dry overnight.

The fit of most joints was so precise that there was little need for the crack-and-gap filler provided with the kit. But we mixed up a small batch just to check it out. It works fine, as long as you remove any excess with a sharp blade after about 15 minutes, and then sand down the filled area about an hour later. After some final finish sanding to remove a few glue spots, we were ready to stain and finish the rest of the assembly.

Both the stain and varnish were in paste form, which made it easy to control the depth of color. We wiped on two coats of stain and followed with two coats (four on the top and shelf) of the clear varnish. After allowing for overnight drying between each application of varnish, we rubbed down the surface with 4/0 steel wool, and achieved a satin-smooth finish. The final steps were to secure the top and the shelf to the frame with the Z clips provided and attach the brass hardware to the drawer front. Needless to say, constructing and finishing this table was an altogether satisfying experience.

The Bartley Collection, Ltd.
29060 Airpark Drive
Easton, MD 21601
Phone: 1-800-BARTLEY
Product literature: free catalog

This graceful reproduction of a nineteenth-century half-round table is easy to assemble and, at less than $130, makes a perfect starter kit. Use it wherever you need a table close to a wall. *(Cohasset Colonials)*

Cohasset Colonials by Hagerty

Cohasset Colonials, the oldest manufacturer of reproductions of Early American furniture in kit form, offers an authentic link with the past, with true copies of museum originals and outstanding pieces in private collections that are "practical" for reproduction. From a single table or chair to an entire room, Cohasset Colonials lets you capture the feeling of the colonial period at reasonable cost and with the pride of accomplishment that comes with the assembling and hand-finishing of any piece of period furniture.

With limited materials and machinery, early American furniture makers could not indulge in elaborate designs. This led to styles and construction techniques that produced furniture simple and sturdy enough to survive far longer than could have been anticipated. Indeed, with its graceful proportions and sheer simplicity, much of that furniture is as modern today as it was two hundred or more years ago. Because they are both fundamental and functional, it's no wonder that Cohasset Colonials reproductions have such wide appeal.

Offered as both kits and pieces hand-finished by its own craftsmen, Cohasset Colonials reproductions are not only faithful to the dimensions of the original but employ the same joinery techniques and, when possible, are made from the same species of wood as the original. The exactness of many of these pieces is such that unscrupulous antique dealers have assembled kits and, after artificially

Cohasset Colonials' best-selling reproduction is its birdcage Windsor side chair, kit-priced at $169. The original, which dates to around 1800, is at Harrison Gray Otis House, Boston. *(Cohasset Colonials)*

"aging" the finished pieces, passed them off to the unsuspecting as the real thing. One way Cohasset Colonials tries to guard against this is to brand each piece with its mark, along with the name of the institution or collection holding the original.

Cohasset Colonials offers some fifty different pieces of Early American furniture in kit form, at prices ranging from $39, for a Shaker candle sconce, to $799, for a Shaker chest of drawers. The majority of its kits are priced at under $200, with a dozen of these costing less than $100. The catalog is a bit heavy on tables, but there's also a good selection of chairs, beds, blanket chests, and benches. Its best-selling reproduction is a c.1800 birdcage Windsor side chair, kit-priced at $169. Also very popular, and a perfect starter kit, is its elegantly proportioned round-top Shaker table. The c.1820 original of this reproduction sold at auction for $55,000. You can have a faithful copy for just $89.

The woods most used in Cohasset Colonials kits are hard maple, clear pine, and Pennsylvania cherry. The kits, just like the original pieces, are designed for simple assembly. The

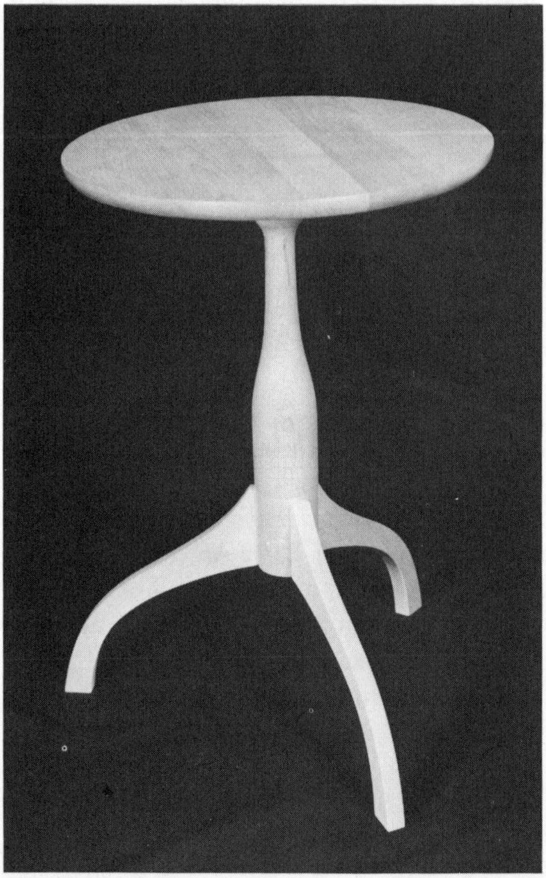

The original of this simple, yet elegantly proportioned Shaker table, or candlestand, sold at auction for $55,000! The kit price is $89. Stain it or paint it in authentic Shaker colors. *(Cohasset Colonials)*

instructions included in each kit are written for those who have never used woodworking tools and who know nothing about wood finishing. With the exception of a few common household tools, everything else you need to assemble and finish the piece is included in the kit: touch-up sandpaper, glue, rope for clamping, and stain.

Although stained finishes are popular today, painted furniture was more common in colonial America. Much of the furniture made for seventeenth- and eighteenth-century New England homes used a combination of woods and was painted or stained for a more uniform look. Paint was also used as a preservative, to extend the useful life of the item. Today, at auction, you'll pay a much higher price for an antique in its original paint than you would for a similar item that someone thought would look better stripped down to the bare wood.

Cohasset Colonials offers a good selection of its own specially formulated paints that duplicate the distinctive colors of early milk-based paints. These paints have an unusually high pigment content and will usually cover in one coat and dry to a hard, durable finish. While each Cohasset Colonials furniture kit comes with a can of its special light stain, which wipes on simply and evenly for a professional-looking result even if you are a beginner, you may prefer to add a coat of Cohasset Colonials stain tone, which, when applied over a coat of the light stain, gives a darker finish.

Most orders for its kits are shipped within three days of receipt. However, some Cohasset Colonials kits are offered or introduced as special cuttings, and might or might not be offered again at some later date. Special cuttings are announced well in advance in its quarterly catalog. If you haven't placed your order before the cutoff date, you might have to wait a year or longer until the next cutting is made. Special cuttings usually represent the more elaborate pieces, such as an extendable tavern table, at $349, a kitchen hutch at $449, or a pencil-post bed, at $579. If you plan to buy a specific kit or finished piece, it's always a good idea to call to make sure it's in stock, or to find out when it might become available, before writing out an order.

Graceful, yet sturdy, the all-maple Vermont kitchen table seats four and also provides the perfect surface for preparing food. Adapted from a country Queen Anne original, kit price is $398. *(Cohasset Colonials)*

Cohasset Colonials by Hagerty
Cohasset, MA 02025
Phone: 1-800-228-2389
Product literature: catalog, $3
Showroom: Main Street, Cohasset, MA

Shaker Workshops

The Shaker movement, which got its start in this country as a nine-member religious sect in the 1770s, observed a doctrine of celibacy, common ownership of property, and community living. Its philosophy, as preached by its founder, Mother Ann (Lee), stressed the possibility of perfection here and now. Shakerism reached its peak in the mid-1800s and has since virtually died out, but not without leaving behind a rich legacy of furniture pieces and other objects that are truly inspiring in their originality and purity of design.

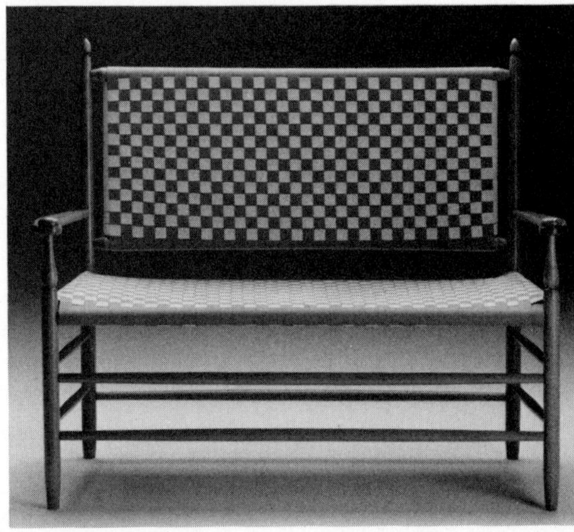

One of the rarest forms of Shaker furniture, there are perhaps no more than half a dozen surviving Mt. Lebanon Settees. The scale of this quality reproduction, kit-priced at $265, is well-suited for today's homes. *(Shaker Workshops)*

The best-known of all Shaker chairs is the infinitely adaptable "slat back" or straight chair. A rock-maple frame and heavy-duty chair tape give Shaker Workshops fine reproduction surprising strength. Kit price is $90. *(Shaker Workshops)*

Serious interest in Shaker objects as collectibles was slow to develop, but they have now become so prized by growing numbers of collectors that prices at auction have been going out of sight. In 1990, Oprah Winfrey, the television personality, paid $220,000 for a Shaker chest-of-drawers. That was ten times the preauction estimate! At that same auction, a simple production candlestand, signed on the bottom by two early-twentieth-century Shaker "sisters," sold for $35,000. More recently, an 1839 sewing desk went for $50,000.

Fortunately, for the many of us who greatly admire the simplicity and versatility of Shaker design, it's possible today to purchase a wide range of well-crafted Shaker reproductions in realistically priced kit form. And no company offers a more exciting collection of Shaker rockers, dining chairs, tables, beds, stools, benches, and other furniture pieces than Shaker Workshops. The Workshops catalog features twenty different chair kits alone. But this shouldn't come as a surprise, since chairs were by far the most popular of the products made in Shaker communities for sale to outsiders, called "the world."

Shaker Workshops chair kits include everything from armchairs and rockers sized for a child to adult dining chairs and rockers in two different sizes and a variety of back styles. There's also a tall weaver's chair, a low-back chair, a chair that can revolve, and a wonderful double armchair that is one of the rarest forms of Shaker furniture. Kit prices range from $65 to $95 for chairs designed for the younger set, to from $90 to $150 for adult-size straight chairs, and to around $175 for rockers. The double armchair, or Mt. Lebanon Settee, has a kit price of $265.

Rock maple is the wood used to fashion the chair frames. All seat bottoms, and some chair backs, are fashioned from fabric tape, usually of two different colors, to form a checkerboard pattern. This is a distinctive feature of Shaker chairs and was considered more practical than seats made of rush or cane, which could dry out and break, as well as pinch flesh and snag clothing.

This robust yet graceful trestle table with carefully pegged breadboard ends was probably first made at the Mt. Lebanon, New York, community. It comes in several lengths, kit-priced from $455 for the 5-foot version. *(Shaker Workshops)*

The Shaker Workshops catalog also features a nice selection of tables, including a robust yet graceful trestle table that can be ordered in 5-, 6-, 7-, or 8-foot lengths (from $455 to $620), a Shaker bed with a rope foundation rather than box spring, in twin, full, and queen sizes (from $375 to $420), as well as inexpensive Shaker pegboard (from $7.15 for 18 inches of board and three pegs). Other decorative, and at the same time functional, items include a key board (on which to organize the household keys), at $35, and a hanging cupboard at $70.

Shaker simplicity makes the assembling of any of these kits child's play. Chair backs come preassembled, and every Shaker Workshops furniture kit comes with sufficient sandpaper and stain. Fabric tape, in the customer's choice of colors, comes with each chair kit. Stain, however, does not suffice for a final finish on rock maple, and you will have to apply a "soft finish" of oil followed by paste wax, or a "hard finish" of shellac or varnish. All were used on original Shaker furniture and are easy to apply.

Shaker Workshops packages a number of wiping stains in different wood tones and also offers a good selection of Stulb's Old Village Paints. It was not at all uncommon for the Shakers to paint furniture pieces, and it was largely because that chest-of-drawers bought by Oprah Winfrey still had its original red paint that it went for such a high price.

Shaker Workshops
P.O. Box 1028
Concord, MA 01742
Phone: 617/646-8985
Product literature: catalog, $1
Showroom: 18 Mill Lane, Arlington, MA

Emperor Clock Company

Long known for its grandfather clocks (see pages 28–34), Emperor has been building an equally solid reputation with its continuing expansion into fine furniture kits, which, like Emperor's tall-case clock kits, are all based on solid 3/4-inch-thick kiln-dried hardwoods. No guesswork is involved with an Emperor furniture kit. All kit pieces are precut and shaped by skilled craftsmen. Nor do you need special tools. All miters, mortises, tenons, and dovetails are precut for ease of assembly. Complete hardware and step-by-step assembly instructions are furnished.

Emperor, which sells both kits and finished pieces at factory-direct prices, offers a number of furniture items in Queen Anne style, including an end table (although end tables, as such, did not exist in the eighteenth century), a drop-leaf oval coffee table, a lowboy, and a secretary desk and hutch. Prices for these items in kit form range from $269 to $799. For the factory-finished pieces, you would pay from $399 to $1,249, or about 50 percent above the kit cost.

Queen Anne has not been one of our favorite furniture styles. Too many manufacturers get heavy-handed when it comes to the shaping of cabriole legs. Their shorter-legged pieces all too often end up with a bowlegged-to-the-max look. With careful proportioning, the designers at Emperor have achieved an aesthetically pleasing look, even with shorter-legged Queen Anne pieces, and we thank them for that.

The Emperor catalog also features three tables and a writing desk in the Chippendale style, three glass-front curio cabinets, and three versions of the traditional butler's table, including a serving cart on casters and a drop-leaf side table. Other noteworthy pieces include Emperor's recently introduced lingerie/-haberdashery chest, which stands tall and has seven drawers of graduated depth from top to bottom. In Pennsylvania cherry, the kit is priced at just under $600.

This elegant and graceful lowboy typifies Emperor's Queen Anne series, which also includes three tables and a secretary desk and hutch. Kit cost in solid Pennsylvania cherry is under $500. *(Emperor Clock Company)*

Another interesting piece is Emperor's captain's chest, with the look and feel of an old sea trunk. Crafted from oak, and with a truncated lid, solid oak handles, and brass-finished hardware, the kit costs $300. We also like Emperor's baby cradle, which has good heirloom potential. Crafted in cherry, the cradle can swing gently or be held stationary by the insertion of a peg. Wooden buttons cover screwheads to enhance the Early American look. Regular price for the cradle kit is $299, but it has been sale-priced at $179.

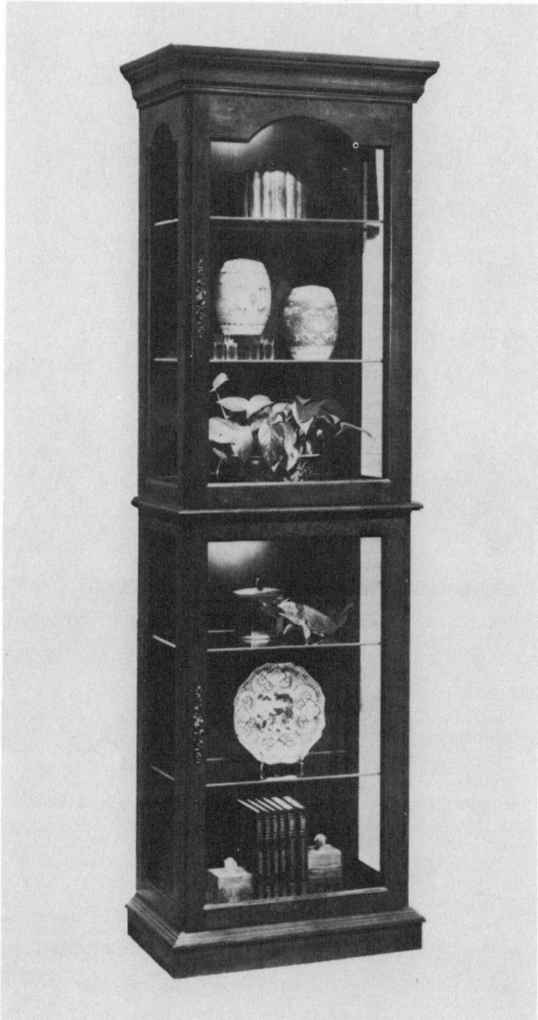

This easy-to-assemble wall curio cabinet creates an illuminated place of honor for your most treasured keepsakes and other collectibles. It stands 72 inches tall. Kit cost is under $400. *(Emperor Clock Company)*

Emperor's kits, like this reproduction of a classic Chippendale writing desk, with laser-cut fretwork, make it easy to own furniture reminiscent of the beauty and grace of earlier times. *(Emperor Clock Company)*

With Emperor, once you get on the mailing list, you're in line for a steady flow of brochures announcing "sales" and "special introductory offers." Emperor's Queen Anne sofa-table kit in Pennsylvania cherry was recently offered at $269 (regular price, $379). The Queen Anne secretary desk and hutch was offered at a $250 discount from the regular price of $799. Best deal of all was a 50-percent discount (down from $399 to $199) for a classic Chippendale foyer table. With Emperor, if you've got your sights set on a particular kit, it could pay to be patient.

Emperor Clock Company
Emperor Industrial Park
Fairhope, AL 36532
Phone: 1-800-642-0011
Product literature: catalog, $1
Showrooms: Emperor Industrial Park, Fairhope, AL; Lake Park, GA, on I-75

Craftsman's Corner

Classic oak furniture and accessories in traditional turn-of-the-century styling is the specialty of Craftsman's Corner. Its extensive product line not only includes something for every room in the home but companion pieces as well. You'd have no problem putting together an all-oak bedroom suite or even an office ensemble from the Craftsman's Corner catalog. There are quality kits here for everything from nostalgic accent pieces like an old-time ice box and a mirrored coat rack to a lyreback commode and a selection of four rolltop desks.

No special skills or tools are needed for assembling most of the furniture kits offered by Craftsman's Corner. We would except its signature piece, the Classic Oak Double-Pedestal Roll-Top Desk, which has long been Craftsman's most popular item and serves handsomely as an American classic in thousands of homes and offices. Our exception applies to the need for a minimum of five bar clamps, two that can span the desk's 54-inch width, and recognition that a *power* finishing sander has it all over hand-sanding when it comes to this much expanse of wood.

All wood parts for the 54-inch-wide desk are precut, including miters, mortises, and tenons, and sanded and coded to correspond to the illustrated assembly guide. All hardware, drawer pulls, and other materials needed to construct the desk are included in the package. The pull-down rolltop or tambour comes fully assembled, ready to be slid into pregrooved slots in the top section. Since the back is fully finished and the desk sits on casters, it is a freestanding piece and can be placed for use almost anywhere. Price of the desk kit is $699. The desk, which boasts ten drawers

Treat yourself to the desk you've always wanted. Craftsman's double-pedestal oak rolltop (kit price, $699) serves handsomely as an American classic in thousands of homes and offices. *(Craftsman's Corner)*

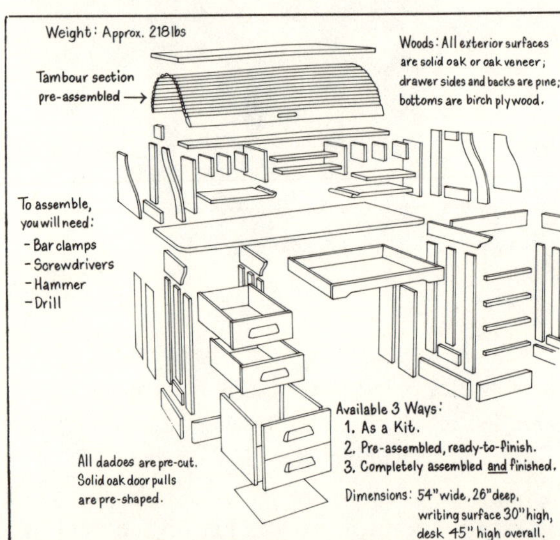

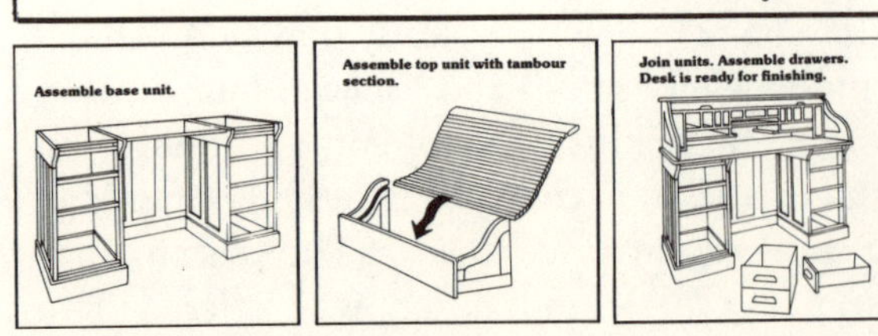

and fourteen pigeonholes, to hold everything from file folders to paper clips, can also be ordered as an assembled, finished unit for $1,667.

Perfectly matched companion pieces to the desk, all in sturdy oak, include a swivel chair ($359), two- and three-drawer file cabinets ($249 and $329, respectively), and glass-door barrister's bookcases (from $230 for a two-shelf unit).

Craftsman's other rolltops include a space-saving, 41-inch-wide single-pedestal model ($479); the Elegant Roll-Top ($400), which is also a 41-inch-wide space-saver, but more delicate in design; and the 28-inch-wide Compact Upright ($350), which sets the desk unit, with a pull-out writing surface, atop three full-width drawers.

One of our favorite Craftsman's items, a perfect starter kit, is a solid-oak rolltop bread box, featuring a preassembled tambour and a pull-out breadboard. Authentically detailed, including white porcelain knobs and oak buttons for a pegged-construction look, the bread box (kit price, $79.95) can add that special touch to a country kitchen.

A perfect starter kit, Craftsman's family-size oak rolltop bread box (kit price, $79.95) is authentically detailed. It features a preassembled tambour and a pull-out breadboard. *(Craftsman's Corner)*

Craftsman's classic oak hutch (kit price, $559) is distinguished by hand-soldered leaded-glass door panels, accented with diamonds of amber glass. All of the period-perfect hardware is included. *(Craftsman's Corner)*

Craftsman's Classic Oak Hutch (kit price, $559) is also worth noting. The hutch is distinguished by hand-soldered leaded-glass door panels, accented with diamonds of amber glass. Period-perfect hardware is included. Overall height is 79 inches, but the base and hutch top are constructed as separate units for easier handling.

If you're looking for a nostalgic accent piece with easygoing charm, Craftsman's Oak Dry Sink fits in anywhere. With two big top drawers and a lower storage cabinet, it makes an enviable bar or an out-of-the-ordinary stand on which to display a few house plants. Kit price is $219.95. Craftsman's Cedar-Lined Chest ($297) is equally versatile. The aroma of cedar wafts from the upper blanket storage section of this handsome chest, which features three side-by-side lower drawers with antiqued brass pulls. With its finished back, it makes a stunning coffee table. Or use it as a charming padded window seat.

Craftsman's Corner
4012 N.E. 14th Street, P.O. Box AP
Des Moines, IA 50302-0556
Phone: 1-800-373-1101
Product literature: free catalog

American Moulding & Millwork

In many of the larger cities, there are stores that specialize in selling assembled, unpainted furniture. From simple bookcases, desks, and chests-of-drawers, the ready-to-finish selections have been expanded over the years to include just about any item of furniture that can be made in mostly inexpensive pine. For a long time, however, a lot of us who wanted it just didn't have easy access to unpainted pine furniture. It loses a lot of its attraction if you have to haul the clumsier pieces home yourself or pay the freight to have the furniture delivered from some distant city.

But then, in the early 1970s, American Forest Products (now American Moulding & Millwork), and other manufacturers of assembled, unpainted furniture, saw the way to reach the considerable untapped market for inexpensive, ready-to-finish furniture. They began supplying both large and small outlets with unfinished furniture in kit form. Unlike assembled furniture, kits don't take up much storage and display space, and they're being carried by an ever-increasing number of retailers, home centers in particular.

By offering well-designed pine pieces that don't cost a lot of money, American Moulding has long been number one in unfinished-furniture sales. Since its founding more than thirty-five years ago, this company, in its several

Constructed from solid pine and steeped in the Shaker tradition, this American classic bench could serve equally well within the home or as a delightful accent in a garden or patio setting. *(American Moulding & Millwork)*

Dovetailing and other fine joinery make assembly of this full-size rolltop desk both easy and structurally sound. At $189 or less for the kit, it can be a most satisfying project. *(American Moulding & Millwork)*

incarnations, has sold millions of pieces of pine furniture. It is with such recent innovations as its Heritage Edition pieces based on Shaker designs that American Moulding maintains its position in a hotly competitive industry.

American Moulding offers an interesting variety of pieces in its several lines of ready-to-assemble furniture. Its Country Pine Traditions includes a four-drawer chest styled after the English bachelor's chest, a spacious six-drawer dresser in the same style, a great-looking armoire that can function either as a dresser/storage chest or as an audio/visual center housing the TV and other electronics, and a full-size single-pedestal rolltop desk with sliding tambour. The hard-to-beat prices (suggested retail) for these truly handsome items range from $119 for the smaller chest of drawers to $189 for the rolltop desk. All of these pieces are of a craftsmanship usually found only in finer, prefinished furniture. The designs are both classic and contemporary, with the precut pieces fashioned from solid, thick-cut western pine.

From its Country Reflections series, American Moulding's buffet cabinet and china hutch are a matched pair that together stand 6 feet tall and could be the centerpiece for a change in decor. *(American Moulding & Millwork)*

It's in American Moulding's Country Reflections line that you'll find the Shaker-styled Heritage Edition pieces. These include a deacon's bench that could serve equally well within the home or as a delightful accent to any garden or patio setting, a nifty jelly cupboard, a double-door country pantry, and an exceptional hutch/buffet combination. Suggested retail prices range from $69 for the bench to $229 for the hutch/buffet.

Unlike much of the inexpensive furniture being sold by home centers and other retailers today, no particleboard or staples are used by American Moulding. Assembly of these kits is made easy with the well-illustrated, step-by-step instructions included in each carry-home package. Finishing tips are also provided.

The pine used in American Moulding kits is all top-grade lumber and can be finished to suit virtually any decor. For a glossy look, you might try a tung-oil varnish. This can be brushed or wiped on, and will require up to four coats, allowing overnight drying between coats. For a country or authentic Shaker look, you might try Stulb's Old Village Paints, which can be ordered by the pint or quart from Shaker Workshops (see page 13). Other treatments might call for the use of stencils or bright enamels. As well-designed as they are, it's not as though you're attempting to replicate a museum piece here.

Among the retailers that carry American Moulding & Millwork kits are Ames, Bradlees, Caldor, Central Hardware, Fred Meyer, Handy Andy, Home Club, and Meijer. You might want to shop around, since prices often are discounted by individual retailers.

American Moulding & Millwork Company
2801 West Lane, P.O. Box 8220
Stockton, CA 95208
Phone: 1-800-344-3237

Outer Banks Pine Products

Freestanding colonial corner cabinets that you assemble and finish yourself are the specialty of Outer Banks Pine Products. Offered in a choice of eight styles copied from museum

Outer Banks 88½-inch-tall Delaware corner cabinet is copied from a museum piece. The lower doors have richly raised panels, and the top doors split the gothic arch. Kit price is $486. *(Outer Banks Pine Products)*

Inspired by an early 1700s Connecticut original, the 82-inch-tall Londonderry is a quality piece worthy of gracing the finest homes. It's also available with split doors. Kit price is $409. *(Outer Banks Pine Products)*

pieces, the handsome corner cabinets are made from 100-percent clear pine, with the precut components sanded smooth on all sides. When purchased in kit form, front frames and glazed, raised-panel doors are fully assembled. The rest of the assembly can be easily completed following the instructions packed with each cabinet. All necessary hardware is included.

Overall cabinet heights, whether with a simple squared top, a crown molding, or a classic bonnet top, range from 72 to 88½ inches. All of the cabinet designs include a double-door lower section and a multilite upper door or doors. (*Lite* is the glaziers' term for the smaller panes of glass that make up a window.) Scalloped-edge shelves, fluted pilasters, and finished backs are among the true-to-the-period details. With their classic design, and finished to complement the decor, any of these cabinets could add a touch of grace to even the most formal dining room. Yet they're also suited to modern demands and utility. Whether the finish is paint, stain, or varnish, light sanding between coats is recommended. Kit prices start at $285.

Outer Banks Pine Products
P.O. Box 9003
Lester, PA 19113
Phone: 215/534-1234
Product literature: brochure, $1

CLOCKS

Good clocks that qualify as home furnishings aren't cheap. But build your own timekeeper from a quality kit purchased at factory-direct prices, and you'll have created a cherished possession for much less than a comparable finished clock would cost, and have the satisfaction, with each tick or chime, of knowing that it likely will long survive you and be passed down from generation to generation as a family heirloom.

One of the most handsome accessories a home can boast is a grandfather or tall-case clock. Building your own grandfather—or grandmother—clock is not nearly so formidable a task as you might think. Clock-kit manufacturers have been in the business of turning out heirloom-quality reproductions a long time and offer kits designed for everyone from the novice woodworker to the master craftsman. With tall-case clock kits, the movements and dials always come fully assembled, and most of the really hard work on the case is done for you at the factory.

Some clock-kit companies sell only precut kits. Others offer everything from plans and true do-it-yourself kits that require the clock builder to cut most of the case parts from raw lumber, to easy-to-assemble precuts and finished timepieces. The choice is yours. Needless to say, the more you can do yourself, the less expensive the final product.

For greater resonance, and also to minimize warping and shrinking, the wood used in most tall-clock cases is solid 3/4-inch kiln-dried hardwood, with oak, cherry, Honduras mahogany, and black walnut the woods most commonly used. These four woods are renowned among cabinetmakers for their durability and dramatic grains. They look truly magnificent when carefully finished.

Precut kits, with all the case parts cut to size, and machined for saw kerfs, rabbets, and moritises and tenons, ready for assembly, are the popular way to go in clock building. When even the screw holes are predrilled, there's little chance of error in the assembly. A fully precut kit usually can be assembled with a few simple hand tools. Phillips-head screws, with two crossed slots, have become common in clock kits, and you'll likely need both a conventional and a Phillips screwdriver, plus a hammer and a simple web or band clamp or two. If screw holes have not been predrilled, you'll find a variable-speed electric drill and wood-screw pilot bits most useful for easing the way and cutting countersinks.

To show you exactly what's involved in constructing a grandfather clock from a precut kit, we've included a photo sequence (see pages 30–33) covering our assembly of America's most popular (in number of units built) grandfather clock. With fifty-four wood parts, the 74-inch-tall clock represents about 50 hours of work, which is in the upper range for precut tall-case kits. With predrilling and a little more preassembly, as offered in several more-expensive kits from other manufacturers, a similarly sized and styled clock might take as little as 30 hours of your time, or a couple of weekends.

With sizes ranging from 71 to 87½ inches, and styles ranging from Shaker simple to the height of elegance, there are models suitable for modest apartments as well as baronial halls. There is no fixed dividing line between grandmother and grandfather clocks. Grandmothers usually are somewhat shorter versions of grandfathers offered by the same manufacturer. With some manufacturers, though, grandmothers represent those tall-case clocks that are less than 6 feet tall. For convenience here, we'll just call them all grandfather clocks, so named from an 1878 song, "My Grandfather's Clock," written by H. C. Work.

Prices for precut case kits, which generally do not include glass, start at around $200 and go to over $1,200. That price includes all wood parts, screws, hinges, and other hardware, but does not include the dial and movement, finishing materials, or shipping charges. Most clock-kit manufacturers use United Parcel Service rather than motor freight to deliver the components, which, with case parts, dial, movement, and finishing materials, can represent up to ten separate packages, totaling between 100 and 200 or more pounds. You pay the delivery charges.

The dial is the focal point of the clock. Its selection, where there is a choice, should be considered carefully. There are both high-arch and low-arch dials as well as square dials. Arched dials are traditional, however, with either a globe labeled *tempus fugit* ("time flies") decorating the semicircular portion of the dial or a moving-moon wheel, geared to the clock movement, filling the space. Moving-moon wheels, which cost more than the nonmechanical dials, reveal portions of the moon to correspond with the current lunar phase. Dials may be deeply etched or embossed with decorative scrollwork, and the numerals may be either Arabic or Roman. Dials, usually solid brass, often are included in the price of the movement. When priced separately, they can cost anywhere from $19.95 to $195.

The slow, stately rhythm of the pendulum and the clear, melodious phrases of chimes ringing out each quarter hour make the clock almost a living presence. Every time you hear it chime, you'll be pleasantly reminded that *you* built it. Grandfather clock movements, nearly all of which are made in Germany, have escalated in price due to the decline of the dollar against the Deutschemark. But weight-driven Westminster movements, including chime rods, weights, brass weight shells, and pendulum, plus the dial, can still be purchased for under $250. With two sets of four balanced and musically tuned rods, Westminster chiming movements play a tune inspired by a phrase from Handel's composition "I Know That My Redeemer Liveth"—with four notes on the quarter hour, eight on the half hour, twelve on the three-quarter hour, and sixteen on the hour—and additionally count each hour with a Big Ben tone.

More elaborate chiming movements feature triple rod chimes and include a selection lever for the playing of Westminster, St. Michael, or Whittington chimes. An option such as triple rod chimes might add $200 to the base price of the movement. Westminster tubular chimes might add $800, while triple tubular chimes could add as much as $1,000. The tone of tubular chimes or "bells"—hollow chrome tubes that hang free in the back of the clock and are played by small hammers hitting the tubes—is often mellower and softer than that of chime rods.

While the bottom price for a grandfather clock kit, including precut case parts, dial, and Westminster chiming movement, might come to around $425, the least expensive packages offered by most of the kit manufacturers start at closer to $1,000. The components for the clock that we assembled, with the exposed case parts in black walnut, the most expensive of the popular hardwoods, came to $509, plus shipping. This price does not include glass or any finishing materials. A few kit manufacturers include glass as part of the kit, or offer beveled-glass sets as an option, but it's usually left to the kit buyer to obtain glass from a local source. There's too much risk that the larger glass panels, especially flat glass, will be cracked in shipment.

If you haven't had any wood-finishing experience involving stains and varnishes, we recommend that you purchase a finishing kit if one is offered. These kits generally include a more than adequate supply of finishing sandpaper, steel wool, stain in choice of tones, and varnish or lacquer. We did a little experimenting in the finishing of our clock and would have had sad results had we continued with the varnish stain (which stains and varnishes in one operation) we purchased from a mass-merchandiser. Fortunately, we tested the product on an interior hidden clock-case surface, and left off when we saw the disappointing results. We tried some other finishes and settled on a Sapolin wood stain in walnut, which we brushed on and then ragged off, to leave just a light, even coating of the stain, since we did not want to hide any of the grain of the black walnut. We then applied Sapolin satin-finish polyurethane clear coating and wiped this to leave just a hint of sheen and the thinnest of coats.

Do not look for shortcuts if you want quality results. Extreme care must also be taken when using white glue to join case parts, or plastic wood compounds to cover punched-in nail heads (brads may be used to secure some glued parts) or fill cracks in joints that don't quite meet. Any glue smears or excess filler not removed by sanding or other means will become more pronounced with staining. All exposed surfaces must be patiently and meticulously sanded to prepare them for a beautiful hand-rubbed finish.

For a final touch, you might want to add a small engraved brass plate identifying yourself as the maker of the clock, along with the year of completion. These can be ordered, with the engraving, from a number of clock-kit manufacturers.

When you're finished, and you've adjusted the clock on its levelers, pulled up the weights, and started the pendulum, you'll have more than just another clock. The craftsmanship that you have contributed will become a special part of family history. Furthermore, you'll have a possession that will likely increase in value. Investors have been bidding up the prices of good clocks, and even the one that you handcraft from a kit for $500 or $1,000 could be worth twice that tomorrow. Just go out and price a few tall-case clocks in the better furniture stores!

As popular as grandfather clocks are, there also has been a strong revival of interest in the old schoolhouse-style clocks, which were throwaways when schools were modernized, and the regulator calendar clocks that were common in railroad stations, saloons, and general stores in the mid-1800s. They're today's chic decorator accents.

Most of the clock-kit manufacturers include school clocks and regulators among the smaller clock kits in their catalogs, often with a choice of quartz (battery) or key-wind movements. The school clocks, which count the hours and strike the half hour, typically feature an octagonal frame, while the regulators take many shapes, in both long-case and short-case versions. These kits come complete with dial, glass, and movement.

There also are authentic replicas of everything from Shaker wall clocks to elegant early European designs inspired by rare museum pieces. With the range of smaller clocks, many of heirloom quality, offered in precut kit form, there should be something here for every home.

No special tools are needed with most of these smaller clocks. About all you need is household glue and a screwdriver. A person of average mechanical ability should be able to put a wall or mantel clock together from a kit with no problem. The usual recommendation for finishing, unlike that for the tall-case clocks, is to stain all pieces before assembly and then apply the final finish after the case is assembled.

With most clock-kit sources, orders can be placed by telephone (often to a toll-free number) and charged to one of the major credit cards.

Mason & Sullivan

For over forty years, Mason & Sullivan has been offering beginning woodworkers as well as master craftsmen an unusually wide range of do-it-yourself clock projects. The company's current catalog includes more than thirty clocks that can be ordered as precut kits complete with movement and dial. With many of these clocks, if you have the tools and the experience, you can order a components-only kit, which includes blueprint, all moldings, turnings, and hardware, but no flat lumber. You can also order a hardware-only kit, or simply the blueprint. With these alternative choices, you would then select an appropriate movement and dial from the selections shown elsewhere in the catalog. Most Mason & Sullivan clocks are also available fully assembled and finished.

Every Mason & Sullivan clock project is scaled to reflect two different levels of skill. Skill Level 1 covers the easiest projects, with all parts precut and most holes drilled. Apart from some sanding, little or no woodworking skill is required. Assembly is with the use of simple hand tools and a web or similar clamp. Most Skill Level 1 clock projects can be completed in a single evening. Skill Level 2 kits also come with all parts precut, but call for more skill in assembling and finishing. The additional tools needed would likely include an electric drill and possibly a few more clamps.

Good basic woodworking skills are necessary if you elect to go with a components-only kit. A table saw is helpful, and you may need a router. The ability to read plans, good woodworking skills, a table saw, and general shop equipment are prerequisites if choosing either of the other two options.

The easiest Mason & Sullivan kits to assemble are its Miniature Clock Collectibles, three unique designs crafted in solid ash or mahogany. You simply sand and finish the case, and then insert the movement. The most expensive of the three designs is the Luna, with a classic tambour-styled case and featuring a calendar-moonphase quartz movement. Kit price for the clock, which stands less than 2½ inches tall, is $34.95. The other two clocks are taller, 4½- and 5-inch pedestal-style timepieces, kit-priced at $19.95.

There are at least ten other clocks, including a Dutch mantel clock with a delft tile dial, a wall clock with a moving-moon dial, and a harbor master's clock designed expressly for East coast tide watchers, kit-priced at under $100.

Mason & Sullivan's three precut tall-case clocks mark the other extreme, in both size and price. These clocks range in size from a 71-inch Shaker Tall to an 84-inch Classic Style Grandfather. The Shaker Tall was inspired by a rare clock case at the Shaker Museum in Old Chatham, New York. The reproduction, in cherry, is true to the original

Inspired by a rare clock case by Erastus Rude (1774–1820) that's now at the Shaker Museum in Old Chatham, New York, this Shaker tall-clock reproduction is true to the original in all but size. *(Mason & Sullivan)*

in all but size. The complete kit, including dial and Westminster chime movement with weights and weight shells, but no glass, is priced at $899.

The case kit for the Classic Style Grandfather, also in cherry, costs $999. Mason & Sullivan leaves it to the customer to choose dial and movement. The least expensive movement for this clock would be an 8-day chain-driven Westminster chime movement at $169, plus $26 for the weights. A high-arch moon dial suitable for this movement would cost $109. Add another $95 for a hand-beveled glass set. That means you could build the

Old school clocks are today's chic decorator accents. This reproduction comes as a precut oak kit, complete with 14-day, spring-driven, half-hour-strike pendulum movement, for $239. *(Mason & Sullivan)*

The classic style and size of Mason & Sullivan's easy-to-assemble Bristol shelf clock, kit-priced at $85, make it a popular choice with novice and experienced clock builders alike. *(Mason & Sullivan)*

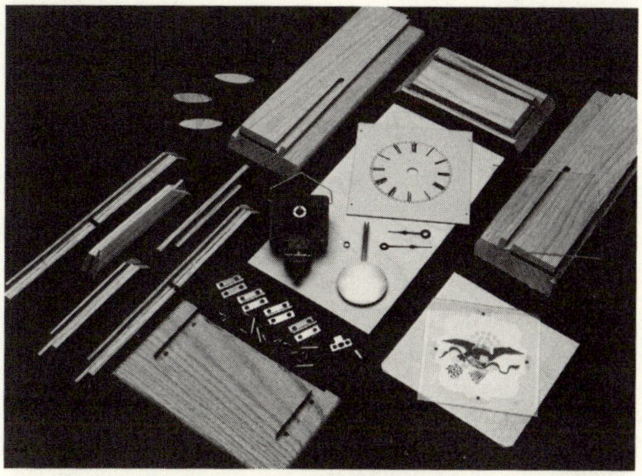

Everything needed to build the Bristol shelf clock, from precut oak parts, hardware, and hand-screened glass to time-only pendulum quartz movement, dial, and instructions, is included in the kit. *(Mason & Sullivan)*

clock for as little as $1,398. For another $58, you could add Whittington and St. Michael chimes. You could also opt for a Keininger 9-tube chime movement at $1,999, complete with dial. This is the movement against which horologists measure all others. In all, there are twelve different movements suitable for this clock.

The Williamstown Colonial Tall Clock is also a 7-footer. The case for this clock is walnut, and the complete kit, including the precut case parts, beveled glass, a triple chime movement with automatic night shutoff, high-arch moon dial, weights, and "mercury" pendulum, costs $1,875.

For the less ambitious and less well-to-do, plenty of other kits here can make equally satisfying projects for first-time as well as experienced clock builders. There are seven different regulators alone. In addition to the more familiar schoolhouse and railroad regulators, there's a coachman's regulator ($539) that is a fine example of a style of clock found in nineteenth-century English and American carriage houses, and a Vienna regulator ($845), often considered the epitome of a clockmaker's craft.

Finishing kits, with ½ pint each of paste stain and paste varnish, together with mixed sandpaper, plastic gloves, and instructions, can be ordered at $19.95. Quarts of wipe-on gel stain are available at $6.95.

Mason & Sullivan's clock catalog also includes an interesting selection of weather instruments in Skill Level 1 kit form. With most of these kits, it's merely a matter of finishing the base or case and then inserting the weather instrument or instruments. As a kit enthusiast, you should also be interested in Mason & Sullivan's *Classics in the Making* catalog ($2), which features scaled reproductions, in kit form, of everything from Leonardo DaVinci's flying machine and the Gutenberg printing press to a live-steam London double-decker bus and ships in bottles.

Mason & Sullivan Company
586 Higgins Crowell Road
West Yarmouth, MA 02673
Phone: 1-800-933-3010
Product literature: *Clock Reference Catalog,* $2.50
Retail stores: 586 Higgins Crowell Road, West Yarmouth, MA; Burlington Mall, Routes 128 & 3, Burlington, MA

Emperor Clock Company

Emperor Clock Company is the world's largest manufacturer of grandfather clocks. Located in Fairhope, Alabama, on the shore of Mobile Bay, Emperor offers both do-it-yourself kits and finished clocks at factory-direct prices. On the one hand, the company would seem to be aptly named; it sells more grandfather clocks than all its competitors combined. On the other hand, as noted in the preceding section, it's also a source of fine furniture kits. Between the two lines, clocks and furniture, it has over four hundred thousand customers worldwide.

Other clock kits are offered in the Emperor catalog, but we'll focus first on Emperor's tall ones. There are at least ten of these regal timepieces, ranging in height from 74 to 87½ inches. In a country where, for many, big is almost always better, it may come as some surprise to learn that it's the "shorty" of the

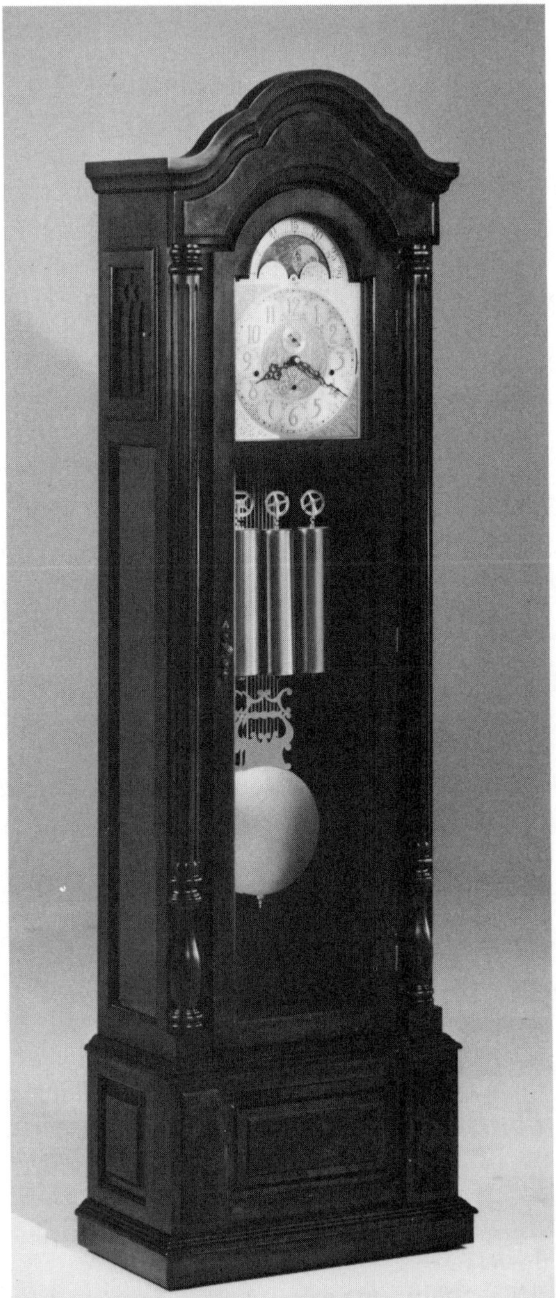

With Emperor's precut kit and detailed instructions, you can build an outstanding piece like the stately 79-inch-tall Windsor even if you've never had any previous experience in woodworking. *(Emperor Clock Company)*

bunch, the Newport, that is America's most popular tall-case clock.

Ordered as a completely finished clock from Emperor, the 74-inch Newport costs $1,595. As a precut do-it-yourself kit, the price (without movement) ranges from $259 in oak to $269 in cherry to $279 in black walnut. Add $219 or $229, respectively, for a Westminster chiming movement with *tempus fugit* or moving-moon dial. A heavy-duty beveled-glass package is available as an $85 option.

As noted in our coverage of Emperor's furniture kits, once you're on this company's mailing list, you'll be on the receiving end of frequent announcements of short-term or seasonal reductions in price. We've seen the case kit for the Newport, in black walnut, priced as low as $199. That's $80 off the regular price. It also means that the clock, bought when the price is right, could be built for about a third of what Emperor is asking for the same clock assembled and finished in its own workrooms.

It's the Newport, Emperor's Model 120-K, in black walnut and with the moving-moon dial and weight-driven Westminster chiming movement (Model 101-M), that we chose to assemble (see pages 30–33) to show you what's involved in constructing a grandfather clock from a typical precut kit.

Emperor's other tall-case clocks include the popular Newbury, which tops out at 77 inches. Unlike the nipped-in-waist design of the Newport, the Newbury and most other Emperor grandfathers are virtually flat-sided. At its narrowest, the Newbury is nearly twice as wide as the Newport at its narrowest. It's a totally different look. The Newbury, however, uses the same movement as the Newport, and its regular kit price is only about $20 more than that for the Newport.

Except for height, the differences between most other models in the line are subtle, frequently reflected in the use of glass panels and grillwork in the clock's sides. The one seriously different design is Emperor's stunning Curio Clock, which, with its flared 36-inch-wide cabinet, is designed to showcase a few of the family's collectibles. Kit price for the Curio Clock in oak is $699. That price, too, has been discounted, at times by as much as $300. The movement for this clock is weight-driven, cable-operated, and has triple chimes. It costs $419. The ten-piece, heavy-duty glass and mirror package, which in this case is not exactly an option, costs $195.

Emperor's top-of-the-line model is the 87½-inch Kingston. Regular kit price for the Kingston, in black walnut, is $799, but it, too, has been discounted, by as much as $200. There's a choice of movements for the Kingston, both of which come with tubular chimes. These movements are priced at $1,099 and $1,299. Don't wait for a discount here, though. We've never seen any of its imported movements discounted by Emperor. The glass package for the Kingston is priced at $150. For the assembled-and-finished version of this magnificent timepiece, with Emperor's top-of-the-line triple chiming tubular movement, you'd have to pay $4,995.

In smaller clocks, Emperor offers an old-fashioned regulator that is an authentic reproduction of the clock that kept the time of day in America's banks, courthouses, and railroad stations through the last half of the nineteenth century. In Appalachian oak and complete with a solid-brass German movement, dial, and accessories, the clock kit costs $289.

Emperor's catalog also features a number of mantel and bracket clocks. Some of these come with triple chiming movements and all make handsome additions wherever you place

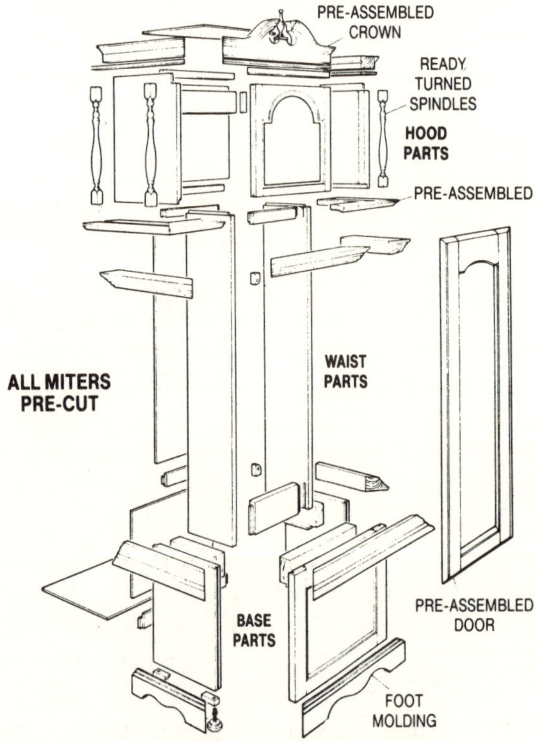

Exploded view of Emperor's model 120-K shows where the case parts go. Construction begins with the base, followed by the waist and hood, and finally fitting the trim and the doors. *(Emperor Clock Company)*

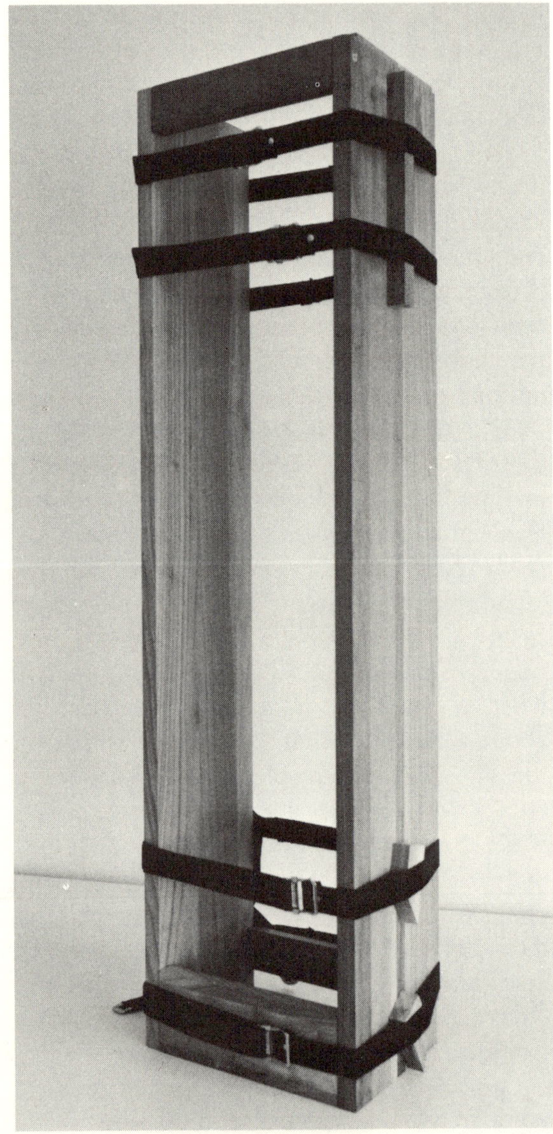

With the rabbets along the rear of the waist sides facing each other to accept the waist back in a later step, the front rails are installed with glue and screws, and then clamped until dry.

(left) Assembly of the base begins with the front and sides. The back panel is in position to ensure later fit. We used a carrying strap to secure the squared-up assembly while the glue dried.

After the hood sides and dial frame have been assembled, with clamp nails driven top and bottom into presawn kerfs, the hood is turned upside down to attach the preassembled waist collar.

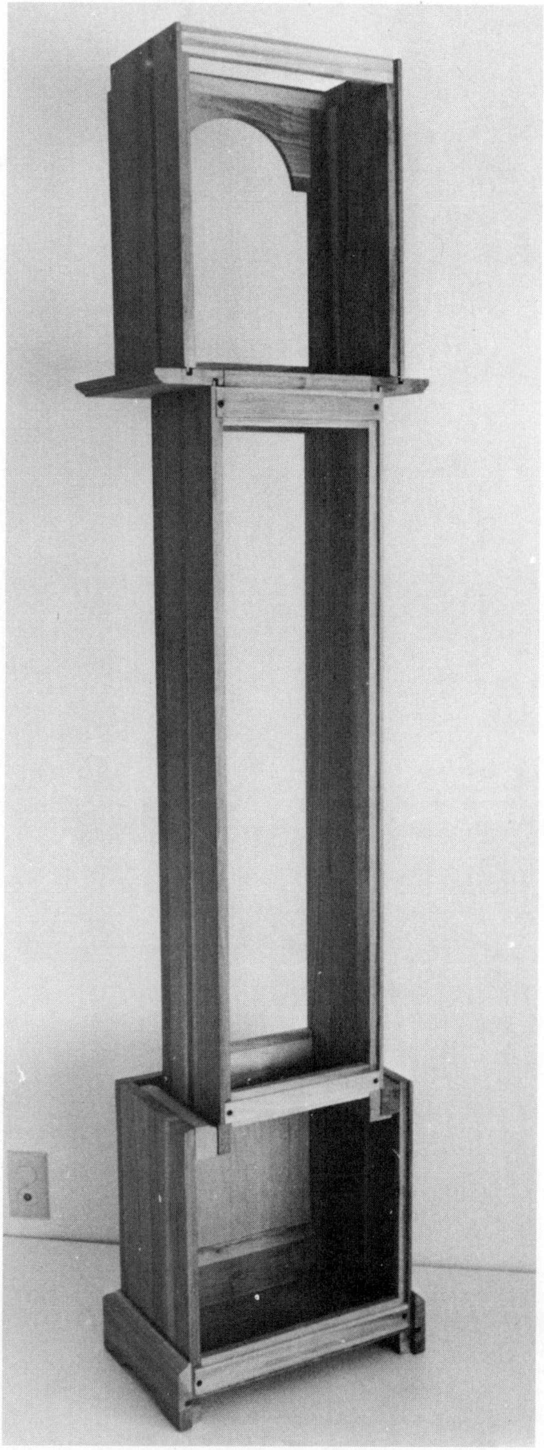

Hood/waist-collar assembly rests on the tenons of the waist sides and is attached with glue and screws driven from inside the hood through the collar and down into the waist sides.

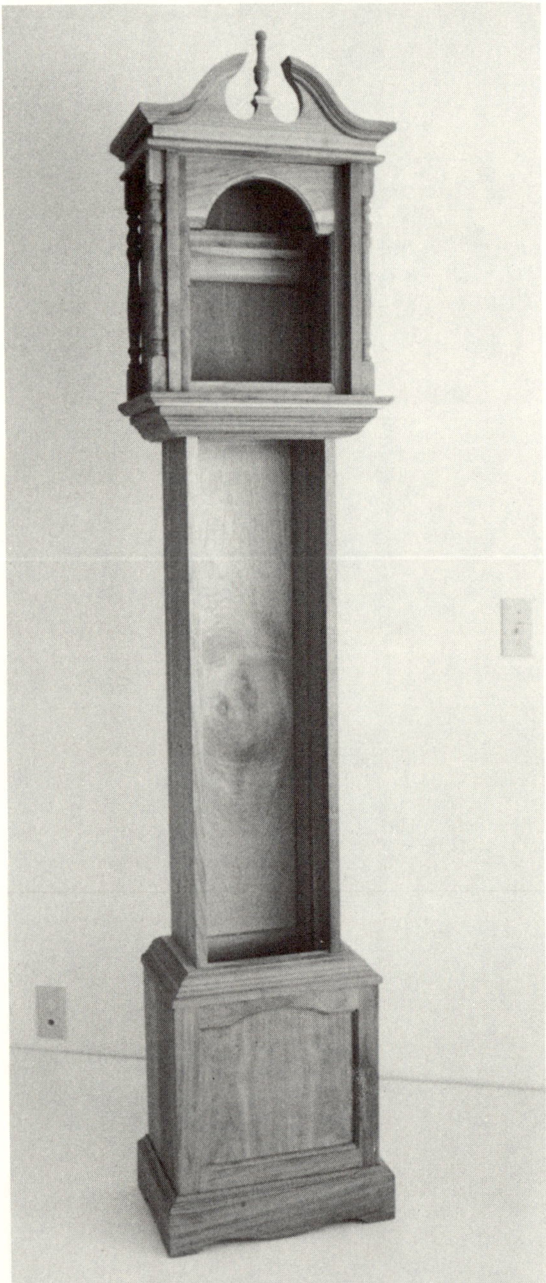

The case assembly is completed with installation of moldings, the hood columns, and fitting the back panels. Doors are hung temporarily and hardware is located to predrill screw holes.

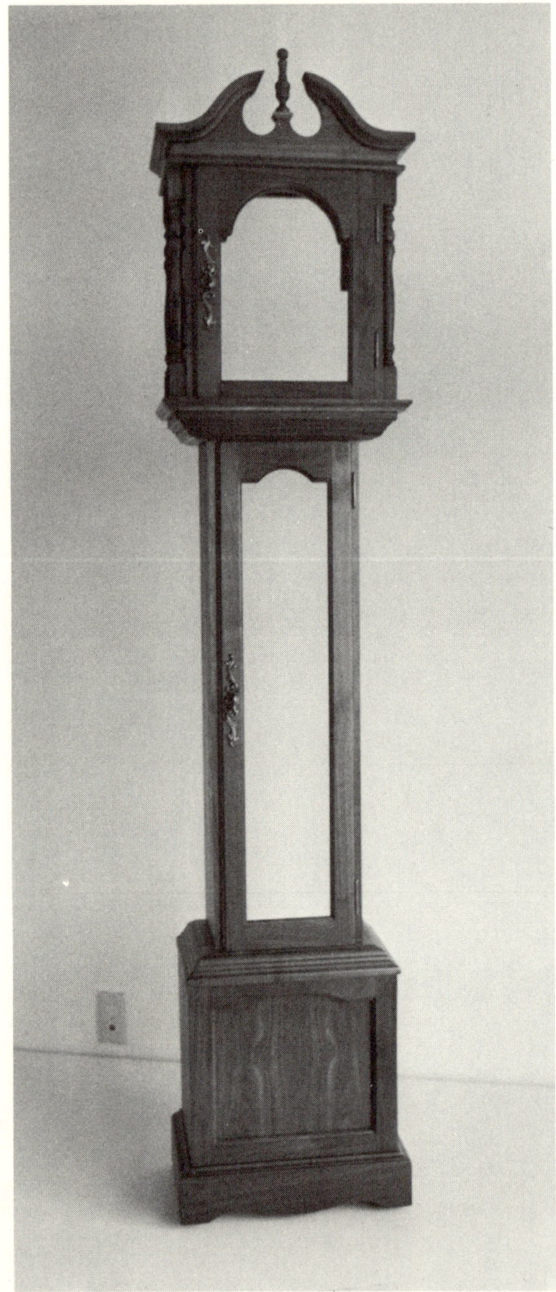

All interior as well as exterior wood surfaces are stained and finished for appearance and to seal the wood from moisture. The doors and hardware are reinstalled after final finishing.

The clock dial and movement come preassembled and mounted on a plywood frame for easy installation. All you have to do is secure the assembly in the hood and mount the chime block.

The chains are released and the brass-encased weights are hung on their respective chains. The pendulum is hung on the suspension arm and started with a gentle nudge to left or right.

With the finial in place, our clock is now complete except for the installation of glass—and perhaps a small, engraved brass nameplate indicating by whom the clock was built and when.

The Berwick mantel clock makes a handsome addition wherever you place it. The kit comes complete with a movement that allows you to choose Westminster, St. Michael, or Whittington chimes. *(Emperor Clock Company)*

them. Bracket clocks come with a handle and in earlier times were moved about the home to wherever they were most needed. Prices for these clocks, in kit form, complete with movement, range from $219 (sale price) to $379. They're easy to build and make a wonderful accent piece for almost any room.

If you are looking to add a touch of Old World charm, you might want to consider one of Emperor's ornate, yet highly functional, wall clocks. Taken from nineteenth-century European designs, these key-wound clocks cost from $169 (sale price) to $449 in kit form, complete with movement, dial, door glass, and accessories. Measuring from 26½ to 34 inches in height, and with a triple chiming movement, one of these might be the answer if you don't have the space or money for a tall-case clock.

You could find clocks like the Gosheim Regulator all over Europe throughout the nineteenth century. The kit comes complete with triple chiming movement, enameled dial, and a silk-screened glass door. *(Emperor Clock Company)*

Emperor Clock Company
Emperor Industrial Park
Fairhope, AL 36532
Phone: 1-800-642-0011
Product literature: catalog, $1
Showrooms: Emperor Industrial Park, Fairhope, AL; Lake Park, GA, on I-75

Kuempel Chime Clock Works

Who should know better how to make grandfather clocks than a group of grandfathers? The Kuempel Chime Clock Works, which has been turning out grandfather clock kits since the mid-1920s, employs retired craftsmen for the most part. There are no assembly lines, quotas, or established working hours at this factory. The craftsmen are allowed to set their own hours, but work with a care that is lacking today at most levels of industry.

There are three ways to build a Kuempel grandfather clock: (1) from a complete Red-i-Kut kit, with joints saw-kerfed, screw holes drilled and countersunk, all parts presanded, numbered, and ready to be glued and screwed together, and the most difficult parts preassembled; (2) from a Kuempel trim kit—you cut the simpler case components from stock lumber and Kuempel supplies the difficult-to-machine parts; and (3) from clock plans, complete with a materials list and full-size patterns, purchased from Kuempel for $11.50.

Kuempel offers nine different tall-case designs, for clocks ranging in size from a 71-inch-tall grandmother, the Spirit of Coventry, to the 86-inch-tall Spirit of America. Complete Red-i-Kut kits for these clocks come in a choice of Pennsylvania cherry, American red oak, or northern-grown walnut, and are priced at $549 for the grandmother and from $699 to $799 for the grandfathers. Case kits come with all necessary hardware and beveled glass.

Three different clock movements are offered for the Spirit of Coventry, priced from $399 to $469. There are eight separate, interchangeable mechanisms, four chain driven and four cable wound, with a choice of chime rods or tubular bells and one or three tunes,

The workings of the stately American Classic, the easiest to build of all the Kuempel tall clocks, can be viewed through the beveled-glass sides as well as through the waist door. With the Mark III chain-driven movement, which plays three tunes on eight tubular bells, and lyre pendulum, kit price is $1,777. *(Kuempel Chime Clock Works)*

available for use with any of the eight grandfathers. These are priced from $629 to $989. The German-made clockworks, which come mounted on a seatboard for easy installation, are unit-priced to include a Kuempel original etched-bronze moving-moon dial, three spun-brass weight shells plus fillings, and mitts for handling the brass. Pendulums are ordered from a selection priced from $45 to $99.

Each moving-moon wheel is individually handpainted by a Kuempel artist, so no two are alike. If you wish, Kuempel will custom paint a "special" moon wheel for you to include a picture of your house, your family crest, or even your family tree.

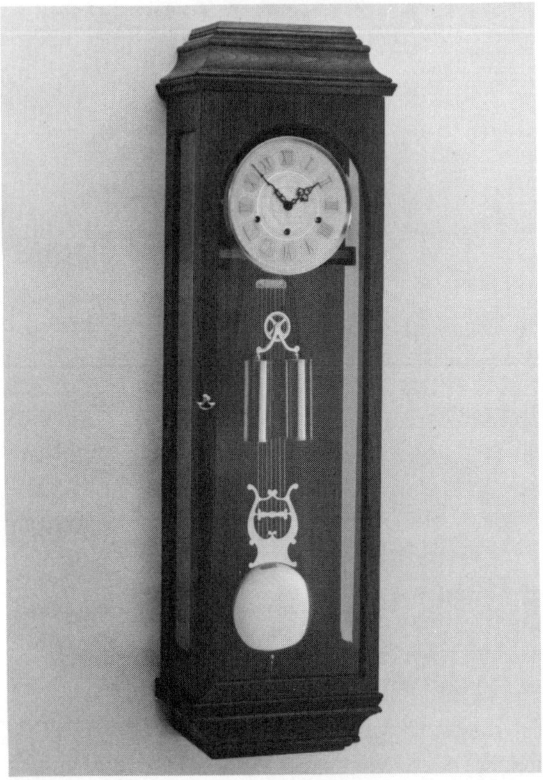

If you haven't the space for a grandfather clock, the Spirit of Liberty wall clock, which plays the beloved Westminster chime on the quarter hour, makes an exceptionally fine alternative. *(Kuempel Chime Clock Works)*

The enduring quality of the Spirit of Ben Franklin parlor clock is evident in everything from the beautifully inlaid hand-made door to the hand-etched glass of the graceful oval side openings. *(Kuempel Chime Clock Works)*

Clock-finishing kits, with everything needed to achieve a professional satin finish, including Velvit oil, stain, light or dark wax, plenty of sandpaper, and scrap wood to practice on, are available at $18.95.

If you don't have the space for a tall-case clock, the Kuempel catalog also features three handsome wall clocks that, while just under 43 inches tall, have a lot in common with the bigger clocks, both in the detailing of their cases and their quality German chime movement. These clocks, an easy weekend project, are kit-priced at $699, complete with movement and hand-wrought bronze lyre pendulum.

There's also an exceptional parlor clock that, with less than a dozen parts, can be assembled in a single evening. It comes in two versions, the Martha Washington, with a coved crown, and the Spirit of Ben Franklin, without the crown. Either way, its quality is enduring. Prices are $325 and $305, respectively, movement and brass dial included.

Any Kuempel clock can also be ordered fully assembled and hand finished, for a premium of from $125 for a parlor clock to $675 for its most expensive tall-case clock.

Kuempel Chime Clock Works & Studio
21195 Minnetonka Boulevard
Excelsior, MN 55331-8605
Phone: 1-800-328-6445
Product literature: catalog, $2
Studio showroom open to the public.

Shaker Workshops

The catalog of Shaker Workshops includes only a single clock kit, but this copy of a timepiece by master clockmaker Isaac N. Youngs (1793–1865), of the New Lebanon, New York, Shaker community, would be hard to match for its understated aestheticism, now so widely admired as Shaker style. Designed for use either on a wall or on a shelf, the original of the clock is now in the collection of the Groton (Massachusetts) Historical Society.

Shaker Workshops' version of the clock, which measures 20½ x 13½ x 4 inches, is authentic up to a point. It features a cherry case with a pine back, a painted wooden dial, and a delicately turned Shaker knob for the door, and even employs four nineteenth-century-style cut nails in the assembly. The clockwork, however, is a quartz-regulated pendulum movement that runs off a C battery.

We liked the looks of the clock so much that we ordered a kit ($115, plus $4.50 shipping and handling; the clock can also be ordered assembled and finished, for $230). The kit includes the precut pieces for the cherry case, trim pieces for the door panel, the back panel, corner splines, glass for the clock face, the clock movement, and all needed hardware. Sandpaper, glue, and a can of Shaker Oil Finish are also included.

Designed for use either on a wall or on a shelf, this copy of a timepiece by Isaac N. Youngs would be hard to match for its understated aestheticism, now so widely admired as Shaker style.

Before beginning the assembly of the clock, we checked the parts in the box against the parts list, and then, as an essential step with this kit, *practiced* assembling the case and the door trim *without applying glue*. Patience and precision are required here to match up the

Patience and precision are required to match up the mitered corners of the case and door trim parts of the Shaker clock. Practice-assembling is an essential first step with this kit.

Assembly of the case calls for working without interruption until all the parts of the case, the clock face included, are glued and then secured with a band clamp for 24 hours' drying.

The special oil finish supplied with the kit enhances the grain of the cherry wood with a natural-looking, low-luster sheen and provides excellent protection. We rubbed on three coats.

After the final finish coat had dried, the door was attached, the glass installed, and the clock movement mounted. All that's left here is to give the pendulum a nudge and close up the back.

mitered corners of the case and door trim parts. Assembly of the case calls for working without interruption until all the parts of the case, the clock face included, are glued and then secured with a web or band clamp for 24 hours' drying.

Most of the wood parts required only a light finish sanding. Some fairly heavy sanding is called for, however, to round the edges of the trim pieces that shape the rectangle at the door bottom, and to prepare the door for hinges. A cabinet file is useful here, and C clamps come in handy when gluing the trim pieces to the door panel. Pilot-drilling is recommended before inserting the screws used to attach the brass hinges and the back panel, and a 1/4-inch drill bit is required for drilling the hole for the doorknob tenon.

For the finish, Shaker Workshops has adapted an early nineteenth-century formula employing linseed oil and other oils in a special turpentine base. The result is a finish that can be applied with a rag to old or new surfaces. It enhances the grain of raw or stained wood with a natural-looking, low-luster sheen and provides excellent protection. We applied three coats and allowed for 24 hours' drying between applications.

After the final oil finish coat had dried, we attached the door, installed the glass, mounted the clock movement, and closed the back. Except for the time allowed for the drying of the glue and the oil finish, it's a weekend project.

Shaker Workshops
P.O. Box 1028
Concord, MA 01742
Phone: 617/646-8985
Product literature: catalog, $1

Klockit

There's no disputing Klockit's claim that it offers the widest selection of clock kits to be found anywhere. The last time we counted, the sixty-four-page Klockit catalog featured at least eighty-nine different kits, along with all the dials and clock movements any clockmaker could possibly want, whether for restoring something old or building something new.

There's a kit here for everyone. But what really sets this company apart are its novelty clocks, which make fun projects for the whole family. These kits include 3/4-inch pine blanks cut to the outline of a baseball glove, football

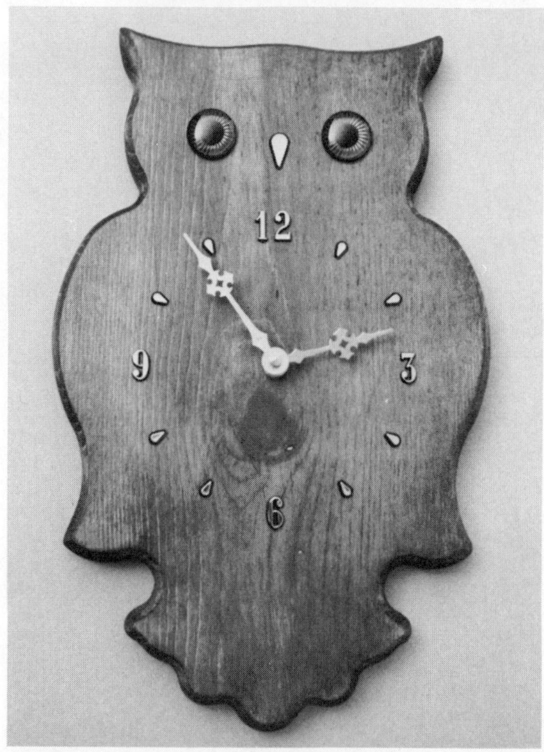

Klockit's inexpensive novelty clocks make fun projects for the whole family. The precut kit for the Owl Clock, which runs off an AA battery, costs just $12.95, quartz movement included.

helmet, cat, pig, teddy bear, owl, or other familiar object, together with a compact quartz movement, clock hands, mounting hardware, and self-adhesive dial or numerals. The blanks have a predrilled shaft hole and are counterbored to hold the movement, which runs off a single AA battery. The kits are priced from $10.95 to $17.95. A paint pattern is included where needed.

Here's an "off-the-wall" project for clockmakers of all ages. The Giant Wooden Wall Watch is nearly 3 feet long! The kit costs $35 and comes complete with an American-made quartz movement.

Glass on three sides plus a rear mirror allows the 14-day, key-wound German movement of this elegant, solid-walnut regulator clock to be viewed from any angle. The kit price is $200. *(Klockit)*

For those looking for an even easier kit, with no painting required, there's Klockit's Americana collection, designed to rekindle fond memories. These kits include a wood cutout with a realistic screen-printed image of a steam locomotive, pot-bellied stove, copper tea kettle, spinning wheel, Model A pickup, or the like. Assembly takes just minutes, and the kits are priced from $18.95 to $29.95.

At the other end of the spectrum, there's Klockit's Vienna Regulator, kit-priced at $499.95. Made from select American cherry wood, the elegant 44-inch wall clock combines classic fluted columns and intricate molding for authenticity and beauty. The kit includes all wood parts, a solid brass Westminster movement with accessories, dial, glass,

The laser-cut kit for the 14-inch-tall Cottage Clock comes with the intricate fretwork already completed. Cutouts in the lower section let you view the swinging pendulum. Kit price is $85. *(Klockit)*

grid pendulum, brass hardware, and complete instructions.

There's also a wide selection of mantel, shelf, and table clocks, including two that allow the intricate mechanism to be viewed from all four sides. Klockit's Mini-Grandfather, at just over 2 feet tall, and a stunning copy of the original, also fits nicely on a tabletop or shelf. It comes with a quartz Westminster movement, for $129.95, a quartz triple-chime movement, for $159.95, or a mechanical Westminster movement, for $229.95.

Other noteworthy Klockit timekeepers would have to include its elaborate fretwork clock projects. Intricately detailed fretwork, once the national woodworking rage, is making a comeback, and Klockit offers both laser precut kits, with the fretwork already completed, as well as patterns for those who would prefer to do the scroll-saw cutting themselves. On perhaps more familiar ground, Klockit offers four versions of the old-time schoolhouse clock, including a compact 16½-inch regulator (kit price, $44.95) that runs on a C battery and keeps precise time with a quality quartz pendulum movement.

Klockit offers many of its kits in two versions: either as a complete kit, or as a plan-and-components kit, which includes movement, dial, hands, hardware, glass, and hard-to-turn wood parts, where applicable. For an example, there's a $75 price difference when Klockit's 24-inch Grandchild's Clock with Westminster chimes (quartz movement) is purchased as a plan-and-components kit and the buyer assumes responsibility for the cherry case and a matching display shelf. The complete kit costs $159.95; the plan-and-components kit, $84.95.

Klockit
P.O. Box 636
Lake Geneva, WI 53147
Phone: 1-800-KLOCKIT
Product literature: free catalog

RUGS

Handcrafted floor coverings have a long history in this country, and two methods of early American rugmaking are experiencing a revival today. The oldest, rug braiding, dates back to the Revolutionary War era. It got its start in New England and reflects the creative ability and thrift of early colonists. The technique employed then was to tear tattered or cast-off woolens and cottons into narrow strips, weave three strips together into a ropelike braid, and then coil and lace a sufficient number of these braided lengths together to create a floor covering.

Little went to waste in New England homes in those early years. And it was practices such as this that gave rise to the term "Yankee ingenuity." The rugs gave comfort underfoot during icy New England winters and could be flipped over once one side began to show too much wear. There was a lot of nostalgia, too, in those rugs. It was still possible for family members to point out what was once a wedding dress, a boy's first suit, a blanket brought from England by an ancestor.

Few homemakers today make clothes for the family and preserve the remnants. Still fewer families have a trunk in the attic (or even an attic, for that matter) reserved for outgrown clothing and other garments no longer in fashion. Most of us would be challenged to find the makings for even a small braided rug within our homes. But there's no reason to let that fact deter you from the satisfaction that can come with a rug that you have crafted yourself. We'll put you onto a source that not only sells the precut materials for making some of the finest braided rugs we've seen but offers kits that include some timesaving construction aids as well.

The other distinctly American rugmaking craft is hooking. Some say hooked rugs also date from the colonial period, but it is more likely that the technique was developed in New England and the Canadian maritimes in the mid-1800s. The craft was pretty much founded on the industrial production of textiles and the spreading use of burlap from India after 1850. The loosely woven material was used for rug backing, and by 1865 stenciled designs on burlap had become available and ensured the future of the craft. Today, the hooked-rug kit, which can make a great project for the entire family, has given this technique of rugmaking unprecedented popularity.

Braid-Aid

Some forty years ago, Roger Griswold revolutionized the art of rug braiding. What had formerly been the most tedious part of braiding was greatly simplified with the introduction of his patented Braid-Aids. These simple metal devices automatically fold each strip of material being braided into a compact tube as it is drawn through, leaving the fingers free to braid tightly and quickly. Previous to his invention, each strip of fabric being braided had to be folded by hand.

Roger and his wife, Marie, founded Braid-Aid to market these and other braiding accessories invented and patented by Roger. Even before Braid-Aids, Roger had come up with the Braidkin, a flattened bodkin shaped for quick and easy insertion between loops of braid during the lacing process. Prior to the Braidkin, all braided rugs were sewn together with needle and thread.

Today, Braid-Aid offers not only accessories that make braiding easy for one and all, but one of the largest selections of woolens in the country for both braiding and hooking. Its Palette coat-weight wool, especially woven for Braid-Aid, is dyed in its own plant to produce thirty-nine Palette wool colors that are exclusive to Braid-Aid.

Braid-Aid is also the leading source of kits for both the beginner and the experienced rug braider. Kits for first-timers include all needed accessories, including three Braid-Aids and a Braidkin, braiding instructions, a row-by-row pattern, and enough precut wool strips to complete an attractive, durable rug. Kits are sized for oval rugs measuring from 2 x 3 to 6 x 9 feet. In Palette wool, the least expensive kit would cost approximately $90. If you are a beginner, however, we recommend that you start out with a kit with less-expensive remnant coat-weight wool. With the accessories, this would cost less than $50 for a 2 x 3 rug, and allow you to get a feel for braiding before you invest in a larger and much more expensive rug kit. Your second kit could be purchased without the accessories. But in Palette wool, even without the Braid-Aids and other devices, expect to pay nearly $450 for a 5 x 7-foot oval-rug kit.

New techniques, colors, and shapes make hand-braided rugs equally at home in contemporary or traditional settings. *(Braid-Aid)*

Braid-Aid doesn't sell hooked-rug kits as such. It does offer a good number of patterns for hooking, printed on burlap, in its 94-page catalog. But almost all of its hooking kits are for smaller items, such as pillow covers, chair seats, and the like. It does offer one interesting kit, Hearts & Cats, that combines hooking and braiding. Price of this complete kit, which makes a 2 x 3-foot oval rug, is $85.

Braid-Aid
460 Washington Street
Pembroke, MA 02359
Phone: 617/825-2560
Product literature: catalog, $4

Shillcraft

If you can't find what you want in its catalog, which features the largest selection of quality latch-hook designs anywhere, Shillcraft's Special Design Department will make up a design to your order. But there should be little need to ask for something special when you have such a wonderful variety of kits from which to choose.

From something as simple as its Alphabet Song design (ABC) to some extraordinary

Orientals, there are rugs here for every setting. With names as widely descriptive as Bonsai, Chinese Roses, Anemones, Unicorn, Montezuma, Rock Island Line, Barney Bear, and Tiger Cubs, for just a few examples, you shouldn't even have to look at a catalog to know why, after more than forty years in business, Shillcraft rates so highly with its customers.

Rugmaking the Shillcraft way is so easy, you can do it while watching TV. Each kit includes a hand-stenciled (for accuracy) canvas, with the design in color; convenient packs of precut, dyed, quality spun yarns—in a choice of 100-percent pure wool or Dupont Orlon acrylic; an easy how-to guide with illustrated instructions; and binding tape, needle, and thread. Shillcraft yarns are the thickest available anywhere and will ensure the longevity of the rug.

Each rug is made up of thousands of pieces of yarn, precut for a deep pile, knotted to the canvas with the aid of a latch hook. The hooking technique begins by placing the piece of yarn—doubled—behind the raised latch on the hook. The hook is then pushed down through one square of the backing and up through the one in front. The two ends of yarn are then bent around the latch onto the hook and the hook is pulled back through the loop. The knot formed is then tightened by pulling gently with the fingers. It doesn't take long to get the hang of it, and even youngsters can become experts in minutes.

The Shillcraft kit for this lovely floral design rug, which measures 36 × 55 inches, costs $169 in wool, $109 in acrylic. *(Shillcraft)*

Convenient packs of cut dyed yarns and a color-coded canvas make rugmaking the Shillcraft way so simple it's a pleasure. *(Shillcraft)*

Shillcraft color-coded rug designs come in a wide range of shapes and sizes. Shapes include circular, semicircular, oval, rectangular, and octagonal. Sizes range up to 60 inches for circular rugs, 60 x 95 for rectangular designs, and 27 x 134 for runners. For a kit for a rug as small as the 18 x 36-inch Alphabet Song, you'll pay $56.50 in wool, $38 in acrylic. Shillcraft's highest kit price is $420 wool, $267 acrylic, for its 60 x 95-inch designs. If there's a design you like but the colors don't match your decor, tell Shillcraft your choices (you've got fifty-eight from which you can choose), and they'll make the color changes free of charge.

Shillcraft
8899 Kelso Drive, P.O. Box 7770
Baltimore, MD 21221-0770
Phone: 301/682-3060
Product literature: catalog, $2

STAINED-GLASS LAMPS

Stained-glass lampshades reached a peak of popularity during the century's first decade. They were the fashionable way to screen the harshness of early bulbs when electric lighting began replacing oil and gas lamps. There were many artisans working in stained glass at the century's turn, but the most prized shades and lamps, expensive though they were, came from the glassworks established around 1900 by Louis Comfort Tiffany, at Corona, New York.

By the 1920s, however, stained glass had fallen from fashion, and Tiffany, despite his fame and early fortune, was bankrupt when he died in 1933. It was another twenty-five years, with the onset of "collecti-mania," before Tiffany lamps were to shine again. When the big auction houses began including Tiffany lamps in their sales catalogs, the bidding wars were on. In 1984, a Tiffany magnolia shade sold at Christie's for $258,000. Just six years later, a lamp signed by Tiffany was sold at auction for $550,000. And now "Tiffany-style" has become virtually synonymous for any lampshade or other lighting fixture, of whatever design, employing stained glass in its construction.

Stained glass has returned to fashion, and a stained-glass shade, whether a Tiffany original or one handcrafted from a kit, makes a stunning accent piece that is bound to draw comments from visitors. With a quality kit and an original Tiffany Studios design, it's possible today, even for the novice glass-worker, to fabricate a stained-glass shade for a table, floor, or hanging lamp that few, other than experts, could tell from a Tiffany original.

Tiffany-style construction, as practiced today, means taking pieces of stained glass cut according to a pattern, edge-wrapping the pieces with copper foil, positioning them with double-faced tape on a plastic form or mold, and then soldering adjacent edges together, forming the leaded lines. Some kits call for the crafter to do the glass cutting; other kits come with the pieces already cut. You save money, of course, when you do the cutting yourself.

Many people have a fear of handling flat glass. But with the proper tools, and once you've acquired the technique, cutting irregular shapes from a sheet of thin glass is no more difficult than working in wood or metal. Actually, the object is to *score* the glass, not cut through it. This is done with the glass cutter held between the first and second fingers, and the thumb pressed against the underside. Properly executed, in one continuous motion, the glass, when tapped, will then fracture cleanly along the score line. Special grozing pliers are used to clean up any "flares" left when cutting curves. Gloves and safety glasses are recommended.

There are petite panel shade kits for beginners, with only a few dozen, mostly straight cuts involved. There are also much more challenging designs that call for hundreds of individual pieces of different colors and textures, and require weeks of painstaking work. Shade diameters range from 7 to 29 inches. Shapes, for which sturdy, reusable sectional and full-size forms are available, include cone, bell, globe, lotus, and wisteria.

Prices of precut kits, such as those offered by Studio Design, Inc. (49 Shark River Road, Neptune, NJ 07753; phone: 1-800-526-2356; catalog, $3), start at around $55 for an 80-piece, 8-inch-diameter shade and go as high as $425 for an 850-piece, classic 28-inch-diameter Spreading Cherry Tree Dome. Uncut glass packages can cost quite a bit less. Appropriately styled bases and the other hardware needed to complete any of their lamp designs are also available from most glass suppliers.

When totaling up the cost of a stained-glass lamp, don't overlook the cost of the tools and such other essentials as copper foil and solder. Dry-wheel glass cutters cost as little as $3. Carbide steel cutters, which last five times longer, cost around $13. But our recommendation would be for a self-oiling cutter, at around $20. You'll also need grozing pliers. These cost from $5 to $10. A good 60-80-watt soldering iron will add another $25 to $50 to your shopping list. Throw in safety glasses, copper foil, a foil wrapper, flux, solder, and other essentials, and you'd do just as well to purchase a glasswork kit for $75 to $130.

The WordenSystem of lamp crafting makes it possible even for novice glassworkers to fabricate a shade for a table, floor, or hanging lamp that few could tell from a Tiffany original. *(H. L. Worden Co.)*

H. L. Worden Company

First introduced in 1974, the WordenSystem of Tiffany-style lamp crafting features reusable lamp forms and matching pattern packets. The forms, only one of which is needed to make a lamp, are molded of thick styrofoam and are surprisingly durable. The system offers twelve different sectional forms and five different full-form shapes, on which more than 160 turn-of-the-century lamp designs can be built. Forms and pattern packets are purchased separately, but every form package includes a design series key, showing all the patterns available for that size and shape form.

The pattern packets include Worden MagicStrips, self-adhesive templates with the lamp design printed on them. Following molded reference lines, these are attached to the curved surface of the form as a guide for the placement of each piece of glass that will make up the shade. With sectional molds, pattern repeats are simply soldered together and slid around the form until the shade is completed. MagicStrips are removable, allowing reuse of the form for other lamp designs.

Each pattern packet also contains glass-cutting patterns that compensate for the curve of the form, a color key with helpful glass selections, and a 16-page instruction manual that details how to score, break, foil, and solder. Pattern packet prices start at $10; forms are priced from $10 to $38.

Glass packages for WordenSystem lamp designs are made up in two grades. The A-Mix Pak contains mostly hand-rolled Uroboros and Bullseye mottled and streaky type glass.

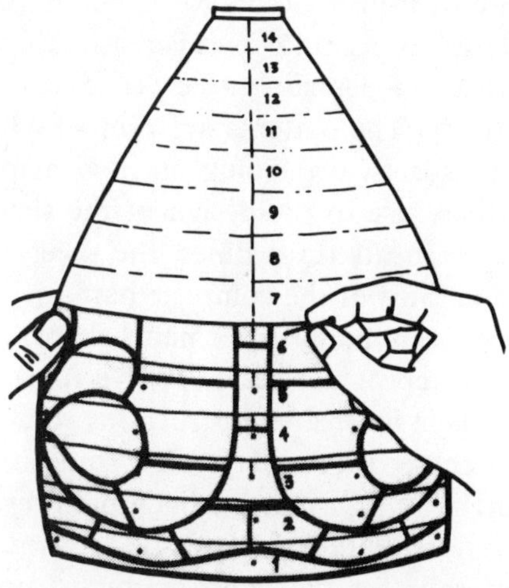

The WordenSystem uses MagicStrips to transfer the lamp design to the curved surface of the form, as a guide for the placement of each piece of glass that will go to make up the lampshade. (H. L. Worden Co.)

All of Worden's Tiffany reproductions are available only in this grade to ensure the best possible match in color and texture to the Tiffany original. Its B-Mix Pak contains some Uroboros and Bullseye glass, but mostly machine-rolled Kokomo-type glass. A-Mix Paks are priced from $18 for a 7-inch-diameter Art Nouveau cone to $185 for an 18-inch diameter reproduction of a Tiffany wisteria shade. B-Mix Pak prices start at $14 for the Art Nouveau cone and go to $136 for Worden's 24-inch-diameter, bell-shape Bird Lamp. The A-Mix Pak for the Bird Lamp costs $173. Beginners are advised to start with a B Pak.

All Worden GlasPaks contain extra glass so the customer can match color flow and density in each repeat of the design and spread the pattern on the glass for easy scoring and breaking.

The WordenSystem lampshade pattern packets and forms are sold worldwide. Look for them in stained-glass shops and hobby stores that carry stained-glass supplies. The company offers a technical assistance hotline.

Outlay for glass, pattern packet, and the reusable full form on which this 246-piece, 16-inch baroque shade is constructed would come to $131 with A-Mix glass, $111 with the B-Mix Pak. *(H. L. Worden Co.)*

Country Tulips, with six repeats and 450 glass pieces, can be constructed as a 20-inch-diameter shade, band and crown included, for an outlay of $174 A-Mix, $139 with the B-Mix glass pack. *(H. L. Worden Co.)*

H. L. Worden Company
118 Main Street, P.O. Box 519,
Granger, WA 98932
Phone: 1-800-541-1103
Product literature: catalog, $1

Whittemore-Durgin

Whittemore-Durgin's stained-glass lampshade construction system also employs sturdy, reusable lampshade forms and offers a wide variety of paper patterns from which to choose, including fifteen adapted from Tiffany Studios designs. The big attraction here, though, are Whittemore-Durgin's bargain-priced glass packages, which cost from $24.95 to $30.95, for more than one hundred different lampshade designs.

The styrene forms on which these lampshades are constructed are all full 360-degree shapes and cost from $21.95 to $23.95. There are no bell, lotus, or wisteria shapes in the Whittemore-Durgin catalog, however, which some Tiffany lamp fanciers may find disappointing. While there are five shade diameters, there are only three different shapes: 16- and 20-inch globes, 18- and 22-inch cones, and a 15-inch panel shade form.

Whittemore-Durgin's reproduction of Tiffany Studios' glorious Grape Trellis Shade is constructed with more than nine hundred pieces of glass and achieves the authentic coloration of the original. *(Whittemore-Durgin)*

Whittemore-Durgin's lampshade patterns, which cost from 75 cents to $2.80, are color-coded and as accurate as it is reasonably possible to be in depicting a curved design on a flat surface. The patterns are printed on various size sheets, depending on how many repeats there are to the design of the shade. If a design repeats three times, the sheet shows only one-third of the complete pattern. Those 75-centers are patterns for panel shapes, most of which repeat the design twelve times.

Complete illustrated instructions come with the purchase of every styrene form, but none are offered with the lampshade patterns. The Whittemore-Durgin lampshade construction system calls for the lamp crafter to prepare three complete copies, with repeats, of the *entire* pattern needed to cover the lamp form. One of the patterns will be taped to the form, another will be cut into stencils, and the third will be used as a locator for assembling the glass after it has been cut. The original color-coded pattern is saved for use as a reference.

If a photocopier is used to make those reproductions, be sure to check copy sizes against the original. The machines are often set up to produce a slightly smaller print of the item being copied. One alternative here would be to purchase as many additional patterns as are needed from Whittemore-Durgin and piece together those three complete patterns using "originals." For a design with three repeats, say, you would order ten patterns, which, at the most, would cost $24.50 (since there's a discount for ten or more).

One piece of each color glass in every glass package is marked with the number corresponding to the color code on the pattern sheet. After the lampshade has been completed, there should be a good supply of glass left for other uses. Based on the cost of the glass pack, form, and pattern, "kit prices" for

1. **Prepare three complete pattern copies from heavy paper.**
2. **Place one complete pattern copy onto the form.**

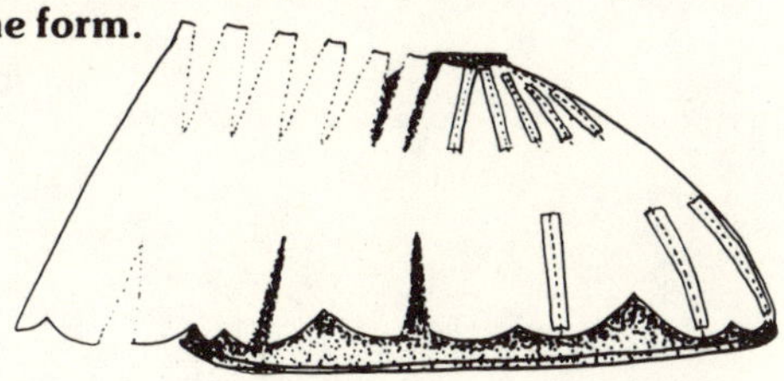

3. **After cutting the glass, wrap the pieces in foil and attach them to the pattern with double faced tape.**

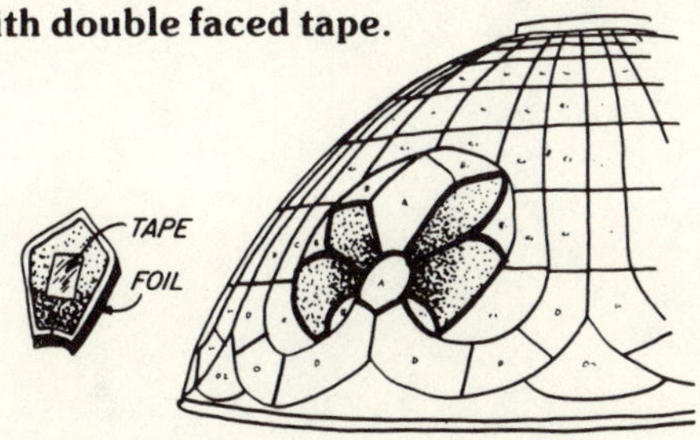

4. **Complete and solder the entire shade.**
5. **Lift the shade and pattern off the form.**
6. **Detach the paper pattern from inside the shade.**

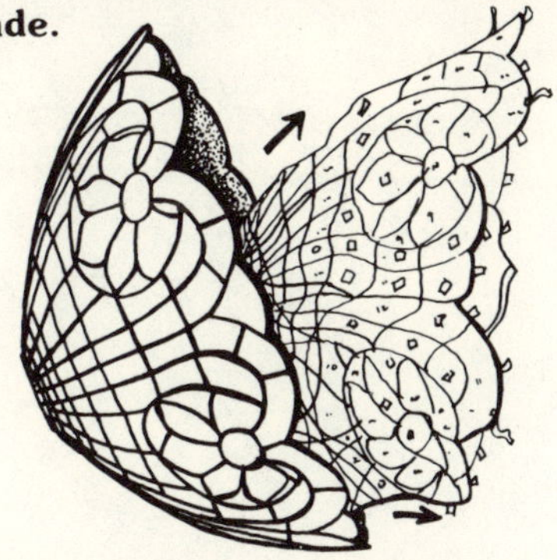

7. **Solder the inside of the shade.**

Whittemore-Durgin's stained-glass lampshade construction system employs sturdy, reusable lampshade forms and offers a wide variety of color-coded paper patterns from which to choose. *(Whittemore-Durgin)*

the more than one hundred lampshades featured in the Whittemore-Durgin catalog range from $47.65 to $57.70. The cost of the bronze filigree for dragonfly and poppy shades in the Tiffany style would be extra.

Catalog items are sometimes dropped due to the lack of a reliable supply of the glass required to build it. This was the case with a reproduction of Tiffany Studios' glorious Grape Trellis Shade, which depicts varying ripeness in the grapes and the aging of the leaves. In places where the vines or leaves cross over the trellis, the glass is raised to give an added dimension. The eight-panel shade is constructed with more than nine hundred pieces of glass, and all of the panels are different. Whittemore-Durgin again has ample glass of the right kind to achieve the authentic coloration of the original, and this challenging project is back in the catalog. The reproduction package costs $219.95 and includes all the materials that actually make up the shade, including copper foil, plus a hardware package for wiring and hanging the lamp.

Whittemore-Durgin has also put together a kit containing materials, tools, instruction books, and the other supplies needed to construct as many as six lamps, depending on size and design. The tools and accessories, of course, will allow the glassworker to proceed on to other projects with the purchase of additional glass and foil. The kit, described as The Big Daddy Lampshade Maker, costs $239.

Whittemore-Durgin
P.O. Box 2065
Hanover, MA 02339
Phone: 1-800-262-1790
Product literature: catalog, $2

Home Improvements

SUN ROOMS

Sun rooms have many antecedents, from elegant nineteenth-century English conservatories or winter gardens to early twentieth-century American sun porches and parlors. Their current popularity, however, owes more to the introduction of "solar rooms" as direct-gain, passive solar solutions to the energy crisis of the 1970s. Although they were designed and built originally to use the free and abundant heat of the sun, that's not what has made them one of today's hottest home improvements.

The sun rooms of today have a lot more going for them than simply serving as a heat sink, though this potential is something that should be taken into consideration when fixing on the intended use if you are contemplating such a purchase. The primary reason for adding a sun room today is to enlarge, enclose, or brighten a portion of the home. It's often a less-expensive alternative to the typical stick-built addition and can result in a 90-percent or better return on your investment when you decide to sell your home, as opposed to about a 60-percent return on an ordinary room addition.

Some sun rooms are added to expand the kitchen and provide an informal dining area, others to replace a dormer and brighten an upper-level bedroom or sitting room. But most sun rooms, though perhaps not planned as such, become a family room or second living room and the one to which family members will most often gravitate. By bringing the outdoors in, without any of the attendant discomforts, a sun room lets you bask in winter sunshine, observe the changing seasons at closer hand, soak in a hot tub or drift off to dreamland beneath the moon and the stars.

Sun rooms come in many shapes, sizes, and materials. They can be virtually all glass, or glass on one, two, or three faces. Roofs can be clear or opaque. Walls can be slanted or vertical, with glass from roof to ground or built with a kneewall. The framing system can be aluminum or wood, curved or straight-eaved. Glazing can be tempered, double-glazed, tinted, low-E, or reflective glass; it can also be a transparent or translucent acrylic or other plastic. You have many decisions to make if you are thinking of adding a sun room to your home.

The first decision applies to the intended use of the addition. If it's to be a solar collector, it will have to be properly oriented; does the face of the house to which the room will be added face south or nearly south? How big should it be? Just large enough to bring morning sunshine into a too-dark kitchen or breakfast area? Or big enough to serve most of the family much of the time? If there's one regret heard from homeowners who have added sun rooms, it is that they wish they had added a larger room.

Are there any lot-line restrictions that will limit the depth of the add-on? You're most likely going to need a building permit if you plan to add a sun room. Don't venture beyond the planning stages until you find out what standards you'll be required to meet under local building codes. The sun room will be a permanent part of the house and will have to meet the same wind and snow load requirements of house construction. Building and possibly energy codes will also likely call for the use of tempered or laminated glazing. And don't lose sight of the fact that any apparent improvement to your home can lead to a reassessment and a higher tax bill.

There are more than one hundred sunroom manufacturers, and their products run the gamut in both pricing and quality. You pretty much get what you pay for here, and if you want a structure that is complementary to the design of the house rather than one that sticks out like a sore thumb or is prone to leak, don't base your selection on price alone. Look for a manufacturer with a good track record, one who is able to meet any special requirements you may have in the design of a sun room for your home. Look for one, too, who stands ready to assist you to whatever degree you find necessary with its installation—whether that's on site, or via a step-by-step video or instruction manual.

The least expensive sun rooms are aluminum rather than wood framed, angular rather than curved. The biggest knock against aluminum is that it too easily transmits heat or cold and contributes to condensation problems unless the framing members are designed with a built-in drainage system and integral thermal breaks. Not all manufacturers go to this trouble. To its advantage, extruded-aluminum framing is lightweight and easily handled by the homeowner doing the job without professional help.

Kits for lower-priced sun rooms may also come with less-expensive glazing. Again, though they may be of lighter weight and more easily handled than insulating glass, fiberglass, acrylic, and other plastics might not stand up as well under prolonged sunlight. With some there can be a yellowing problem; others are easily scratched.

For a quality kit, and installation by a professional crew, including foundation, electrical work, and flooring, expect to pay around $80 per square foot, or better than $15,000 for the typical 12 x 16-foot sun room. You can bring it in for quite a bit less if you prepare the foundation (usually concrete slab) yourself and complete the assembly of the sun room without a lot of professional help.

Once the foundation is in place, it doesn't take an experienced two-person crew more than two or three days to install the average sun room. As a reasonably competent do-it-yourselfer tackling the job with one or two helpers, it might take you a day or two longer.

There's a lot more that needs to be said about the selection and installation of sun rooms, as well as how to handle such problems as ventilation and overheating. We'll touch on these in the following pages as we take a close-up look at some sun-room manufacturers who, in our judgment, are among the best in the business.

Lindal Cedar SunRooms

Few manufacturers can match the clean, uncluttered lines of Lindal's cedar SunRooms, which come in both straight- and curved-eave configurations and are remarkably easy to build. Its Lowline and Contemporary straight-eave models offer a wide variety of roof pitches, making it easy to integrate a Lindal SunRoom with an existing roofline. As well-designed and constructed as a piece of fine furniture, these are sun rooms that can en-

A well-planned, well-built sun room, like this clean-lined, straight-eave Lindal design, can add a whole new dimension of luxury and year-round enjoyment to any home. *(Lindal Cedar SunRooms)*

hance any architectural style, whether traditional, contemporary, or modern.

Lindal's standard models are all 9 feet long (three bays) and include two cedar endwalls. Standard depths are 3, 6, 9, and 12 feet. These are modular systems, and the length of any Lindal SunRoom can be extended with the addition of one or more 3-foot bays. Glass endwalls, doors, or windows to replace the standard cedar endwalls are an option.

With a preengineered Lindal SunRoom package, all the critical cutting, drilling, notching, and routing will have been done at the factory. So let's run through a typical Lindal straight-eave SunRoom installation. It begins by bolting a 4x4 subsill onto the prepared slab or insulated floor system. The critical step here is to square up the sun room with the house. A 2x6 sill is next nailed on top of the subsill and a multipiece ledger assembly, by which the sun room will be attached to the house, is nailed or bolted into position.

Front and notched top mullions—cedar 4x4s that serve as the main framing members of the sun room—are bolted together. Tilted into position, they are butted against the sill and anchored in place. Aluminum flashing is then laid over the sill assembly.

Setting the tempered-glass panels, which come sized to fit the framed openings, consists of first applying double-faced glazing tape to the prerouted outer face of the mullions. This not only provides a cushion to support the glass panels but creates an airtight seal between glass and cedar. Setting blocks are next nailed into position along the base and against the eave filler to provide support and a resting place for the front-wall and overhead glass.

Beginning with the overhead panels, glass is laid in place and heat-seal tape is applied to join panels one to another and provide an additional watertight seal. Cedar trim covers the heat-seal tape and is held in place with bronze screws. Front and end walls are trimmed out in the same fashion, with precut and predrilled 1x4 and 1x6 clear cedar caps.

That's pretty much it. To make the installation even easier, Lindal covers it in a step-by-step video and a 60-page installation manual.

If its SunRooms can't provide what you are looking for in a sunspace, you might want to consider one of Lindal's several other sun products. Designed primarily for kitchens but popular in bathrooms, too, Lindal's straight-eave SkyWall sits on a 40-inch base, and insets into the roof pitch to a depth of 3 or

Lindal's cedar-and-glass SunRooms blend beautifully with its cedar homes and rarely look out of place, whether added during the planning stages or as an afterthought. *(Lindal Cedar SunRooms)*

Built atop a kneewall finished to match the house exterior, this five-bay SunCurve unit offers a look and feeling that is more traditional sun parlor than most designs. *(Lindal Cedar SunRooms)*

6 feet. Where overhead glazing might not be desirable, such as in hot climates, you might consider a vertical SunWall, which still gives you a wide panorama of glass, in 4- and 6-foot glass heights.

Or how about a SunDormer, with glass walls and glass overhead, for a second-story bedroom or bathroom? These are available in virtually any size. Lindal also offers the hip-roofed SunCorner, designed to wrap around a corner and open up the home on two sides. These, too, are available in nearly any size.

While Lindal offers fifteen standard models representing the sizes and roof pitches most frequently ordered, any of its sun products can be customized to your exact requirements. And even beyond dimensional changes, there are many Lindal options that can make for a more personalized place in the sun. Most SunRooms, for example, can be ordered to rest on a standard 33-inch base or kneewall, with a 4-foot glass wall above. You can fit an endwall with French doors or a sliding glass door. Awning windows, with screens, can be installed for ventilation purposes.

With overhead glazing, summertime overheating can be a problem nearly everywhere. Even if you have operable windows or doors for cross ventilation, you're almost certainly going to need an exhaust fan or two, preferably with a thermostat. Options in glazing, such as Low-E, Heat Mirror, and reflective glass, can help block the sun's heat, but you may also have to invest in summer shades. These work best when installed on the exterior of the sun-room roof.

Prices of Lindal's standard straight-eave model begin at $3,020, for a unit three bays (9 feet) long and a bit more than 3 feet deep. For a unit 12 feet long and 9 feet deep, base price would be $6,300. Lindal's SunCurve model in the 12 x 9-foot size is package-priced at $9,900, or about 50 percent more than the straight-eave design.

Lindal Cedar SunRooms
4300 South 104th Place, P.O. Box 24426
Seattle, WA 98124
Phone: 1-800-426-0536
Product literature: 28-page plan book, $2

Solar Additions

We don't know of any other sun-room company that provides the services that Solar Additions does. Before they ship one of their preengineered sun rooms, they put it together exactly as it will go up on the customer's home, to ensure that all components fit together like the movement of a fine watch. It is then dismantled in major, but manageable,

Gracefully curved laminated arches are a feature of Solar Additions sun rooms. This Corner Arch design allows family and visitors to enjoy two horizons at once. *(Solar Additions)*

With a well-insulated roof and thermal-pane windows angled to take maximum advantage of the winter sun, a solar room allows you to sit back and enjoy nature in comfort. *(Solar Additions)*

sections for easier, more accurate assembly at the site. They also operate a seven-day-a-week hot line for the benefit of homeowners who elect to go the d-i-y route with a Starter Room package, one of three purchase options.

Solar Additions' basic design calls for a solid, insulated roof and angled, front glass wall. Together, these keep the room cooler in summer, warmer in winter than most other sun-room designs. Special kneewall windows are incorporated to draw in cooling breezes along the floor while warmer air is pushed up and out through other openings. In the colder months, the window angle takes maximum advantage of the sun, and the room has the potential to produce more than enough heat for itself. The surplus can then be moved into the house with a fan system that, in itself, is pretty special.

Every Solar Additions room is a full 12 feet or more from front to back. Lengths start at 13 feet and can be increased in 3- or 4-foot increments. Glazing is double-pane thermal glass and almost 1 inch thick. The beams that support the roof system and into which the glass is inset are either right-angled heart of

With illustrated step-by-step instructions and help only a phone call away, Solar Additions makes it easy for the do-it-yourself homeowner to assemble one of its sun rooms. *(Solar Additions)*

redwood or curving, laminated Southern yellow pine. It's the latter beams, sweeping arches at least 3 inches thick and made up of fifteen or more laminations, that, for us, truly distinguish this manufacturer's sun rooms. The apparent sturdiness and the warmth of so much natural wood can contribute to a feeling of relaxed well-being not always possible in sun rooms less soundly constructed.

Solar Additions sun rooms can be purchased as a complete package, a partial package, or a starter package. All three options include the necessary beams, insulated glass panels, and kneewalls. The complete package also includes side or endwalls, doors and windows if ordered, exterior framing and finishing materials, and the complete roof system. The partial package includes everything except endwalls. The starter package calls for the do-it-yourselfer to purchase all readily available materials from a local lumberyard or home center.

The easy way to go, of course, is with the complete package. No sawing or sanding is required, and the easy-to-manage sections come ready to be bolted into place. The kneewalls come in 4- or 8-foot sections, and even the endwalls come in preassembled 4-foot-wide sections. With all the preassembly, it is possible for two or three people to put one of these sun rooms together in a single weekend. The instruction manual is complete and, as previously noted, help is only a phone call away. Solar Additions is also caring enough to rout recesses into the face of bottom support frames to avoid any finger-mashing when setting in the heavy window panes.

Solar Additions recommends the use of skylights with a roto opener and built-in insect screen in the sun room's roof. Skylights can help immensely in the ventilation process as well as allow you to view the constantly changing overhead show. Another recommended option is a reversible, multispeed fan. The one offered, an English import, is engineered to operate automatically at any preselected room temperature, to move air quietly and efficiently—from sun room to home and vice versa.

While the soaring arch is a feature of many of the sun rooms ordered by Solar Additions customers, there are also some more traditional designs employing straight beams and sloped or vertical glass. The company is also prepared to put together a custom package to suit whatever special requirements you might have, from enclosing both a patio and a second-floor deck to enclosing a swimming pool.

Solar Additions, Inc.
Route 40
Greenwich, NY 12834-9882
Phone: 1-800-833-2300
Product literature: 40-page *Guide to Sun Room Living*, $6

Florian Solariums

Florian has been in business for more than fifty years, which gives it more right than most to refer to its sun rooms as solariums. But don't think for a moment that this is a company that time has passed by. Florian's solariums are probably the most technologically crafted sun rooms available today. The company has led the way with everything from maintenance-free, anodized-aluminum framing and tempered, curved glass to its thermal-break system and Heat Shield insulated glass.

It is a full-line manufacturer, and Florian's aluminum-framed Sierra Room series alone

Curved eaves soften the lines of this Sierra Room installation, creating a warm, inviting atmosphere. Straight eaves might have the edge where a more formal look is preferred. *(Florian Greenhouse, Inc.)*

includes thirty-two stock models, in both curved-eave and straight-eave configurations. And that's before you begin adding panels or bays beyond the standard three. Its Woodhaven Solariums series, in pine or cedar, begins with eight curved-eave and eight straight-eave models. In short, there's a Florian solarium for every style home, whether eighteenth-century colonial or California contemporary. Choose a curved or straight-eave model, a basewall or glass-to-ground unit, bronze or white baked-on enamel finish, laminated pine or laminated cedar. If you order a custom unit, Florian will even assemble and test it in the factory before delivery.

We expressed some caution regarding aluminum-framed sun rooms earlier, but Florian's thermally broken, anodized-aluminum frames are a technological marvel. The Sierra Rooms' intricate aluminum extrusions, multiple gaskets, rubber seals, and exclusive nylon clip assembly eliminate problem-causing metal-to-metal and glass-to-metal contacts. An integrated condensation and water-release track further eliminates interior moisture build-up by automatically weeping condensation out of the unit.

The anodized-aluminum framing for a Sierra Room comes precut, predrilled, and ready to assemble. Argon gas-filled Heat Shield glass is standard and, complete with superior EPDM gasket, the highest grade on the market today, comes ready for installation. The lower heat-conductance characteristics of argon, which is colorless, odorless, and completely harmless, significantly improve the U and R values of Heat Shield.

The wood frame of a Florian Woodhaven solarium is also thermally broken, to cut down on unwanted transfer of heat or cold and to reduce the risk of moisture build-up. As with Sierra Rooms, the framing comes precut, predrilled, and ready to assemble. Heat Shield insulated glass again is standard. All Florian solariums come complete with frame clips and other installation hardware, plus a parts description list and step-by-step assembly guide.

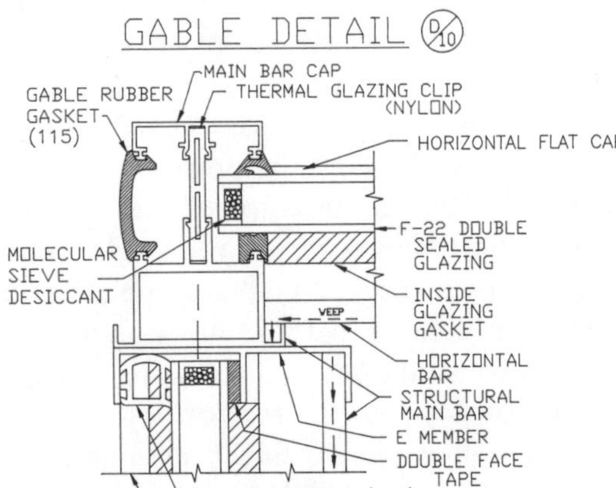

Lest you think that a Sierra Room is simply extruded aluminum and panels of insulating glass, this gable detail should give you a fair idea of the complex technology involved. *(Florian Greenhouse, Inc.)*

The framing for this straight-eave Woodhaven solarium, built with a basewall, is laminated pine. The interior, too, has been finished in pine, for an informal, country look. *(Florian Greenhouse, Inc.)*

Options offered by Florian include bronze-tinted and heat-reflective Heat Shield, access doors, ventilation windows, intake/exhaust fans, and both interior and exterior shading systems. Exterior shading is recommended with aluminum framing. In use, the shades prevent the sun's direct rays from reaching the framework and glazing, cutting down on heat build-up and lowering cooling costs.

Prices of stock, straight-eave Sierra Rooms begin at $2,152 for a unit three bays (8 feet) long and a bit more than 3 feet deep. For a unit 13 feet long and 8 feet deep, base price would be $5,017. A curved-eave Woodhaven solarium in the 13 x 8-foot size is kit-priced at $7,469 in pine, $8,216 in cedar.

Florian Greenhouse, Inc.
64 Airport Road
West Milford, NJ 07480
Phone: 1-800-FLORIAN
Product literature: 32-page catalog, $5

Sun Room Company

Conservatories are very English and have been adding an extra touch of elegance to homes of the well-to-do in Britain since the eighteenth century. Sometimes freestanding but more often an extension of the house, they originally served as garden rooms, for the growing and showing of flowers and other plants. Today, in this country, they're a stunning alternative to the more common sun-room designs.

Sun Room Company is the first American manufacturer to offer conservatories of classic English design. Previously, if you wanted a sunspace in intricate Victorian or Tudor styling, it had to be imported from England. With framing handcrafted from clearheart redwood and double-fastened at every joint for superior stability, Sun Room's conservatories are in a class with the best the English have to offer.

Sun Room's grand-scale Warwick conservatory is indicative of the company's ability to translate each client's dreams into pleasing reality. *(Sun Room Company)*

The company provides a free custom-design service and numerous options that can make a conservatory truly special—from arched grids in the bays and dentil molding at the eave to cresting and finials on the ridge. Copper flashing and a white paint finish are available, too.

Sun Room Company also offers sun rooms of more conventional design, in two standard bay widths, 28 and 34 inches, that can be used in combination to meet any size requirement or design wish. Ten standard depths are offered, ranging from 4 to 16 feet. The framing for these sun rooms, too, is handcrafted from low-maintenance, clearheart redwood, which is a true outdoor wood that actually improves in appearance with age. Glazing is double insulated and nearly 1 inch thick.

Sun rooms are delivered to the site preassembled into the largest possible modules consistent with ease of handling. Preassembled gable ends and front walls, with all joints dadoed, glued, and screwed, assure the integrity of the structure. Precut rafters, prebent flashing, caulking, and sealing materials, with explicit instructions for their use, make installation of the sun rooms practical even for the do-it-yourselfer.

An exclusive feature of these sun rooms is a concealed, continuous Ridge Vent System that allows for circulation of air without being subject to potential mechanical failure. Weather-, insect-, and bird-proof, the system is left in the open mode during the warm months and closed during the cold months to retain sun-warmed air.

Glazing options include 3-lite insulated glass and a variety of glass combinations. Where shading is needed, the manufacturer suggests

(*above*) The rotunda design of the redwood-framed Donegal conservatory gives owners and guests a 270-degree view of the ever-changing world around them. *(Sun Room Company)*

(*below*) The coziness of this Donegal interior suggests the potential for a Sun Room conservatory becoming one of the most popular rooms of the house. *(Sun Room Company)*

Solarium Component Assembly

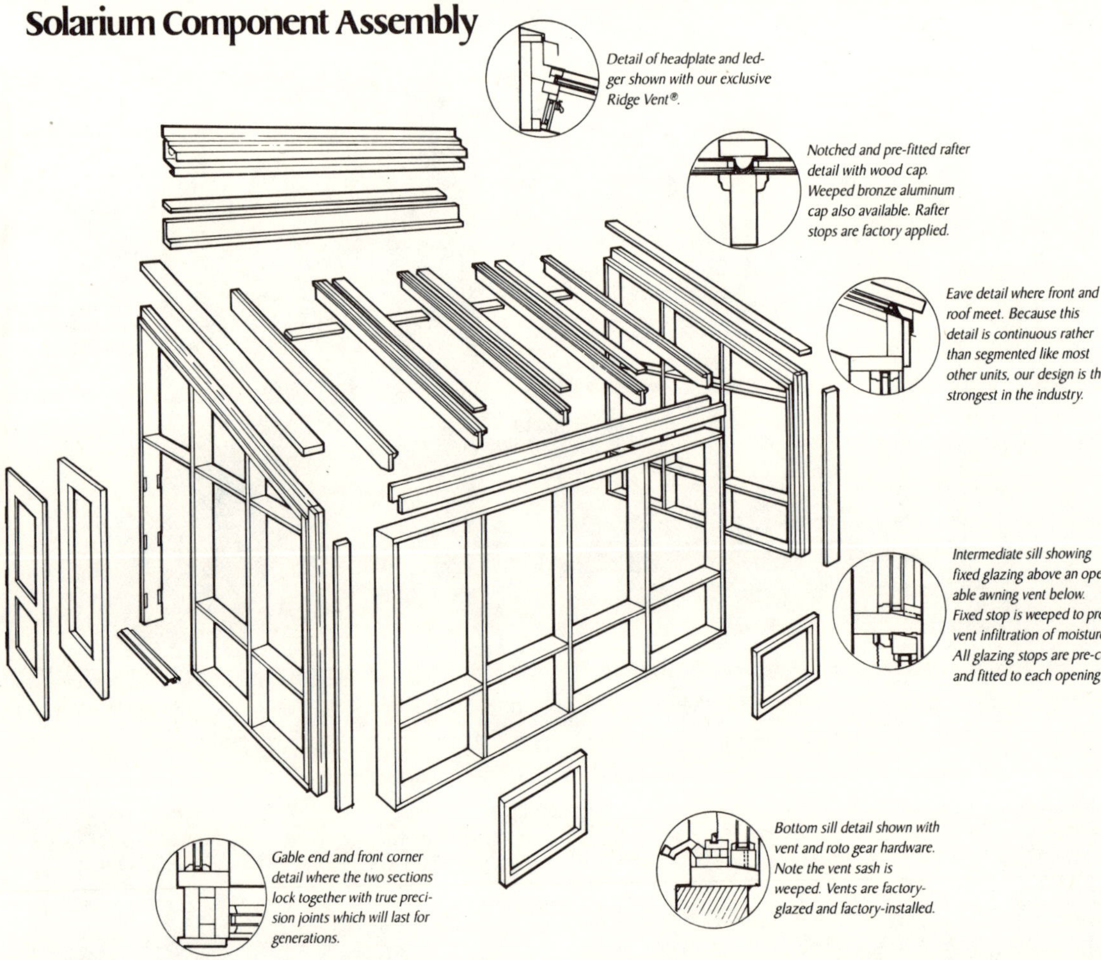

that this can be most effectively achieved through the use of heat-reflective glass or exterior Solar Sun Screens. Sun Room prices begin at $5,765, for a conventional 8 x 4-foot unit. A sun room measuring 13 x 8 feet is base-priced at $11,686. These prices do not include delivery, installation, or taxes.

Sun Room Company, Inc.
322 East Main Street, P.O. Box 301
Leola, PA 17540
Phone: 1-800-462-2737

Sun Room's components for its traditional solariums are delivered preassembled into the largest possible modules consistent with ease of handling. *(Sun Room Company)*

GARDEN WINDOWS

Replace an existing kitchen, bathroom, or other flat window with a bump-out garden window, and you can not only open up a dark, cramped area but provide a year-round place in the sun for the growing of everything from herbs to African violets.

As a remodeling project, installing a garden window is not that difficult. Most units come preassembled or nearly so, and with some you don't even have to remove the existing window. Those units are designed to fit on the exterior wall, outside an opening or a double-hung window. Others, in standard window sizes, are provided with a sleeve or flange and are fitted into the opening left when the old window is removed.

Just as with sun rooms, there are both straight- and curved-eave window greenhouses. Frames run the gamut from all-aluminum to all-vinyl to all-wood and clad-wood. Some come with tilt-up roofs for ventilation, others with slider side windows that raise or lower to allow comfortable cross ventilation. Still others come with a hinged, full-size ventilating sash that swings out from the bottom. Bump-outs typically protrude between 12 and 18 inches.

NorthStar garden windows by Wenco are equipped with insulating glass, and the single-hung flankers are fully operable for ventilation. The divided-lite grilles are offered as an option. *(Wenco Windows)*

Window-box greenhouses, like this one by Sun Room, are unique replacements for "two-dimensional" windows. They can add a new dimension to your kitchen, bathroom, dining room, or den. *(Sun Room Company)*

We've seen bronze-finish, aluminum-framed garden windows priced as low as $186 for a 3 x 4-foot unit. However, we would have second thoughts about hanging one of those on even a mobile home. For a quality unit that can be considered a home *improvement*, be prepared to pay a lot more than that.

There are three companies worth mentioning here. Two of them, Lindal Cedar Sun-Rooms and Sun Room Company, have already been singled out for the quality of their products, in the preceding pages on sun rooms. The other is Wenco Windows (P.O. Box 1248, Mt. Vernon, OH 43050; phone: 614/397-3403), a manufacturer whose only business is window and door products.

Wenco's NorthStar slope-top garden windows are equipped with insulating glass, and the single-hung flankers are fully operable for ventilation and double weatherstripped to keep frigid air out. Frame and sash are constructed from Ponderosa pine and factory finished in a choice of four colors to provide for a low-maintenance interior and a carefree exterior. A rich fruitwood interior stain is also available at no additional cost. The upside of the seatboard is veneered with top-quality oak. The window projection is just under 15 inches, and the unit comes in nine different sizes. It is set up for a standard wall thickness, and screens and sash locks are included. Prices start at around $600.

Profiled wood grilles that give a true divided-lite look and are removable for easy cleaning are a Wenco option. Internal muntins are also an option. Sandwiched between the panes of insulating glass, they *never* need dusting and do not interfere with glass cleaning.

Lindal's "three-dimensional" garden windows are framed to blend with the cedar exterior of Lindal homes. These windows include a mid-height shelf as well as venting side panels that open to catch a summer breeze. Only insulated glass is used, to enhance energy efficiency. A typical Lindal unit, measuring 5 x 3 feet, and 12 inches deep, with glass shelf and screens, would cost $1,070.

Sun Room Company's window-box greenhouse units are constructed with clearheart redwood and double-insulated glazing. These garden windows come with adjustable glass shelves and a removable copper drip pan for the floor of the unit. Redwood knee braces, both functional and decorative, help to bear the weight of the installed unit. Available options include colonial grids, redwood shelves, and solar sun screens. The windows come with a choice of roto-gear venting styles, awning or casement, and can be dimensioned for exterior mounting or to be sleeve-mounted into almost any rough wall opening. There are thirty-six standard configurations, priced form around $850 to $3,000. All necessary hardware, screens, flashing, and caulking are part of the package.

CUPOLAS

While we tend to think of cupolas as a purely decorative feature of colonial-style homes, their early use in this country was as a venting device installed atop barns and other storage facilities to minimize the build-up of gases produced by fermenting grain. Later, they were adapted by the gentry to ventilate the upper reaches of their large homes.

Cupolas also made an ideal perch for a weather vane, a necessity for colonial-era farmers and sailors needing to know the direction of the breeze; by the end of the eighteenth century, the pairing had become a common but distinguishing feature of a good many New England homes and town halls.

But form has since nearly overcome function, and, while few cupolas are installed today without a weather vane, most are added as a "crowning touch" rather than to handle attic heat build-up. Soffit venting combined with ridge venting takes care of that problem in most homes built since the 1950s.

Almost any cupola installed on a roof ridge can be made to serve as a natural ventilator. But their popularity today is as a design statement, and cupolas can be found in shapes, sizes, and roof treatments suitable to many different styles of homes. Most are placed over the main part of the house at the center of the roof. Another popular location is on the roof of an attached garage, centered over the doors. They even come scaled for garden sheds and doghouses, too.

Size can be most important when selecting a cupola to be mounted high up on the ridge of a house. Choose one that's too small, and it'll look like a birdhouse up there. If too large, it can overpower the building. There's a formula to go by here. The base of the cupola should measure *at least* 1 inch for every foot-length of uninterrupted ridge. If in doubt, order the next larger size. An even better idea is to take a cardboard box that closely matches the dimensions of the cupola you're considering, position it where the cupola will go, and then view it from the ground. This will give quick confirmation of how a cupola of the proper size will look on your roof. It also provides an opportunity to measure the pitch of the roof, a necessary first step when ordering a ready-to-mount cupola or installing one that comes as a kit.

Roof pitches vary and, to ensure a proper fit, an accurate measurement of the angle formed at the ridge is essential when notching the base of the cupola. The angle for these cuts can easily be determined by resting a flat stick on each slope of the roof and allowing the sticks to intersect at the peak or ridge line. Pencil lines are then scribed on the sticks where they cross each other. The sticks can then be nailed or clamped together to maintain the angle, which can now be easily transferred to the cardboard box or a paper template, and ultimately to the cupola base.

For a snug fit, even if the cupola is for decoration only, it is recommended that the portion of the ridge cap that would otherwise be covered by the cupola be removed. To secure the cupola, the most common procedure is to nail 2x4 cleats to the roof to correspond to the interior base outline, and then nail through the notched base sides into the cleats.

If the cupola is to be used for ventilation, care must be taken not to cut into rafters or the ridge beam when cutting the opening in the roof. As a general rule, the opening should be 3 inches less all around than the exterior outline of the base. Screening stapled over the opening will keep out insects. Caulking is needed only where the notched sides of the base and the roof meet. The other faces of the base are secured to the roof with nonferrous nails or wood screws driven at an angle squared to the roof and of sufficient length to penetrate the roofing boards.

A cupola is a relatively inexpensive addition that not only can add a touch of elegance but add to a home's value. Most kits cost under $400 and, if a cupola is appropriate to the design of your house, could be an improvement well worth further consideration.

Country Cupolas

With 517 standard models and fifteen different roof styles, Country Cupolas, in northern New Hampshire, offers the largest selection of cupolas to be found anywhere. All models are available fully assembled or as money-saving, do-it-yourself kits. The cupolas are constructed from solid northern white pine. Copper roofs are standard, though some models (those with curved roofs are the exception) can be purchased with an unfinished roof. This might be the choice of someone planning to finish the roof of the cupola to match the roof of the house.

Country Cupolas kits take approximately 4 to 6 hours to assemble and install. They include all needed precut and lightly sanded wood parts, wood glue, fasteners, aluminum screening, detailed assembly instructions, and, if applicable, the copper roof covering. The classically simple Windsor, a four-sided design, is this manufacturer's most popular model. It's available in base sizes ranging from 16 to 36 inches, with overall heights of from 25 to 47 inches. Kit prices, copper roof included, range from $106 to $462.

Country Cupolas offers the largest choice of cupolas of any manufacturer—in shapes, sizes, and roof treatments suitable not just to colonial, but to many different styles of homes. *(Country Cupolas)*

Other four-sided designs include the Pembroke, which has a mansard-style roof, and the Somerset, which has four well-proportioned gables. The Derby is distinguished by a bell-shaped roof, and the Canterbury has a gracefully curved pagoda-style roof. The Derby and the Canterbury represent designs where the unfinished-roof option isn't applicable. The Somersets are the most expensive of Country Cupolas' four-sided models, with kit prices ranging from $165 to $721, copper roof included; with the unfinished-roof option, $119 to $519.

Country Cupolas also offers three six-sided and three eight-sided series, the most expensive of these being the impressive Epsoms, which, with eight sides and a bell-shaped copper roof, are kit-priced from $357 to $1,036. Standard base sizes for all six series range from 20 to 36 inches, and overall heights from 31 to 66½ inches.

Any of this manufacturer's cupolas with a base size of 20 inches or greater can be ordered as a nonventilating bell or lantern cupola. Louvers are omitted for visual effect, and the base is covered with a solid deck to keep out the elements. A decorative low rail outlines the openings with elegant simplicity. The bell is a functional cast-metal farm bell, and the electric lantern is handcrafted in copper. Either option adds from $92 to $207 to the cost of the kit, depending on base size.

Pleasing six-sided symmetry and the flowing curves of a bell-shaped copper roof distinguish this manufacturer's Newmarket series of cupolas, which are kit-priced from about $285. *(Country Cupolas)*

Country Cupolas also offers a nonventilating mini-version of its popular Windsor cupolas. It is designed for use on poolside cabanas, small utility buildings, and garden sheds up to 12 feet long. Kit price of the Mini Windsor, with copper roof, is $77.

Weather vanes to top off its cupolas can be ordered through Country Cupolas' Country Things mail-order catalog. The catalog includes an exceptionally wide choice of ornaments and finish options. Prices for complete weather vanes, suitably proportioned for every cupola size, start at $30. Top-of-the-line Country Things weather vanes, with a full-bodied ornament and scaled for larger cupolas, cost as much as $530. But that's with a 23K gold-leaf finish. With a pewter or gold-bronze finish, those same vanes cost $245.

Early weather vanes have become so prized by collectors of American folk art (in 1990, a splendid 1860s copper and zinc weather vane of a horse and rider was sold at auction for $770,000!) that they've become a target of thieves. The homeowner might be well advised to settle for a less-expensive weather vane rather than one so authentic to the period that it would take an expert to distinguish it from an antique.

Country Cupolas
East Conway, NH 04037
Phone: 603/939-2698
Product literature: free Country Cupolas catalog; free Country Things catalog

Stephenson Cupolas

Designed for easy installation by the do-it-yourselfer and sold through local home centers and lumberyards, Stephenson cupolas come largely preassembled. To make doubly certain that their construction has a sturdiness to last for years under all kinds of weather conditions, wood components are handfitted and treated against decay and insects. The pine base and louvered midsection come primed with a protective oil base that can be painted or stained. Capping it off are a handsome copper or aluminum roof and a hand-cast aluminum weather vane.

Stephenson cupolas by Webb are offered in a variety of styles and sizes. They are attractive in appearance and price, provide attic ventilation, and are easy for the homeowner to install. *(Webb Manufacturing, Inc.)*

There are three series of Stephenson cupolas. The Standard series, for contemporary homes that have colonial features and gently sloping roofs, includes twelve models, with base measurements ranging from 22 to 47 inches. Approximate heights range from 25 to 36 inches.

The six-model Governor series and the four-model President series complement more traditionally designed homes with sharply pitched roofs. The Standards and the Governor series represent four-sided designs. The President series of cupolas are all six-sided models, with pagoda-shaped roofs.

Prices will vary depending on the retailer, but figure on a range of $175 to $350 for the Standard series of cupolas, plus another $45 for a Standard weather vane. Prices on the Governor series start at around $275.

Stephenson Cupolas, Webb Manufacturing, Inc.
Conneaut, OH 44030-0707
Phone: 216/593-1151

Detailed instructions are provided with each Stephenson cupola kit for everything from cutting the roof opening, if called for, to securing the cupola roof and attaching a weather vane. *(Webb Manufacturing, Inc.)*

SPIRAL STAIRS

Back stairs used to be common in most larger homes, but many of us who live in today's smaller multilevel residences don't know the convenience of having a second set of stairs. If you're looking for a short cut, from the kitchen to an upstairs nursery, say, or from a second-story deck down to a patio, spiral stairs can provide one of the most efficient ways we know to conquer space. They can also provide an alternate escape route in case of fire.

Spiral stairs are often used to access sleeping lofts in vacation homes where space is at a premium. They don't take up a lot of space, and their open design doesn't obstruct the view. But that's one of the few situations where we'd want to settle for spiral stairs as the *only* staircase in a bilevel environment.

Dramatic and space-saving though they may be, spiral stairs do have their drawbacks. Some people find the more tightly spiraled stairs sized for installation by the homeowner difficult to negotiate. They can also present problems when it comes to moving large pieces of furniture.

Whether for use in new construction or as a remodeling project, all stairs must meet riser, tread, and other specifications under local building codes. You might have been thinking in terms of a smaller-diameter, less-expensive spiral staircase, when the smallest stair diameter acceptable could be 5 feet. So don't order a kit without first verifying with the building authority that it's up to code.

The smallest spiral stair diameter offered by most kit manufacturers is 3½ feet. While this might be acceptable for a second home or an exterior staircase, we would recommend at least a 4-footer (a diameter of 5 feet would be more comfortable) for use in a primary residence. You also need to know the vertical distance from finished floor to floor when ordering a spiral stair kit. This determines not only the height of the center pole but the number of risers required.

One of the many options offered with spiral stair kits is a choice of spiral direction. This, looking down, can be either clockwise or counterclockwise. You also have a choice of material. Steel frames with wood treads are popular, as are oak frames and carpeted treads. In a home with young children or small pets, extra balusters to fit in between the normal baluster spacing are recommended.

Preparation of an upper, through-the-floor well opening could take some time with a retrofit. Added support beneath the lower floor may also be needed. But the actual installation of the spiral stair shouldn't take two people more than a day.

Prices for spiral stair kits start at under $500.

The Iron Shop

It would be hard to dispute The Iron Shop's claim that it offers the "best buy" in spirals. For almost sixty years, this company has enjoyed a reputation for outstanding design and fabrication of spiral, curved, and floating stairs, and today is the world's leading manufacturer of spiral stair kits and accessories.

An Iron Shop stair kit can be as simple or as elegant as you like. Its least expensive kits, priced from $425, are of all-steel construction with a vinyl handrail. But you can also choose from a full line of accessories to customize the stair to your own particular taste. This might include specifying a brass or oak hand-

Painted to resist rust, a metal spiral stair serves here to provide an inexpensive connection between terrace and patio. *(The Iron Shop)*

rail, oak tread coverings, extra landing rails, and (recommended for households with small children) a landing gate. If for exterior use, the kit can be ordered with a hot-dipped galvanized finish.

The Iron Shop's standard metal spirals are offered in eight diameters, ranging from 3½ to 7 feet. There's also a UBC series, with diameters of 5, 5½, and 6 feet, specially designed to meet the dimensional requirements of BOCA and Uniform Building Codes.

You could always paint those metal stairs if you want to match a particular decor. But

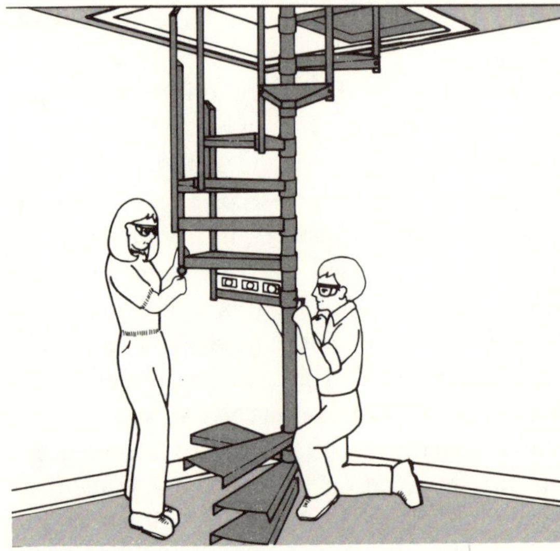

It doesn't take any special skills to assemble an Iron Shop spiral stair kit. Normal assembly time is 4 to 6 hours. *(The Iron Shop)*

then, if you're looking for more warmth and elegance in a spiral stair, The Iron Shop's oak stair kits might be the better choice. These are custom-cut to fit the floor-to-floor height and are priced from $1,575, for an 11-riser, 4-foot-diameter stair kit.

The oak spiral kits are designed for quick and easy assembly. Everything for normal installation is provided, including the top landing, the landing rail, and all hardware. Balusters and landing rail posts can be ordered in colonial, European, and traditional "turnings" as an optional extra to complement a particular decor.

For something extraordinary in a spiral stair, there's The Iron Shop's Victorian One, which has the look of traditional cast iron but weighs only about a third as much. All components, except for the handrail, are solid castings of high-strength aluminum alloy. It is most appropriate in Victorian settings, but it can also be used as a striking contrast in contemporary ones. Its all-aluminum construc-

The Victorian One, shown with a brass handrail, has the look of traditional cast iron but weighs only about a third as much. *(The Iron Shop)*

tion and stainless steel hardware make it uniquely suited for outdoor use, too.

The Victorian One comes in just two diameters, 4 and 5 feet, and is kit-priced from $3,300. The standard kit offers a choice of cast-aluminum or solid-oak tread and landing inserts. The solid-oak inserts are recommended, since the open-grate design of the aluminum inserts could pose a hazard to the wearer of high heels. Among the options: scroll tread ends; brass inserts for treads and landing; brass or oak handrails; and a factory enamel finish in a choice of colors, including "antique" verdigris, the coating that forms like rust on aged copper.

The Iron Shop
400 Reed Road, P.O. Box 547
Broomall, PA 19008
Phone: 1-800-523-7427
Product literature: free color brochure
Showroom/warehouses: Broomall, PA; Pomona, CA; Sarasota, FL; Houston, TX; Chicago, IL; Stamford, CT

Duvinage

America's largest manufacturer of custom stairs, Duvinage has been supplying staircases to some of the best addresses in America since 1895, including the White House and the National Gallery of Art. While much of its output represents fairly elaborate staircases for both commercial and residential applications, Duvinage also produces spiral stair kits for indoor or outdoor installation by the homeowner.

Duvinage preengineered spiral stair kits are available in diameters of 3½, 4, 4½, and 5 feet. Kits include eleven treads and a platform and are adjustable to a maximum height of 9½ feet. All steel parts are painted with a rust-inhibitive forest-green primer, which makes them an ideal stair for outdoor installation. Just follow the step-by-step instructions that come with every Duvinage kit and, in a mat-

Center pole, end caps, and balusters of this striking Duvinage custom spiral staircase are all fabricated of stainless steel. *(Duvinage)*

ter of hours, you'll be walking up and down your own spiral staircase.

The standard Duvinage kit is priced from $530. Optional accessories include three styles of wood tread, a mill-finish aluminum handrail, extra balusters, and additional well rail. The standard tread may also be covered with carpet, but you would have to handle this job yourself or contact others. For best results, Duvinage recommends that carpeting be installed with neoprene cement.

A stair kit modified with extra balusters provides access to a loft in a school building that has been converted to condos. *(Duvinage)*

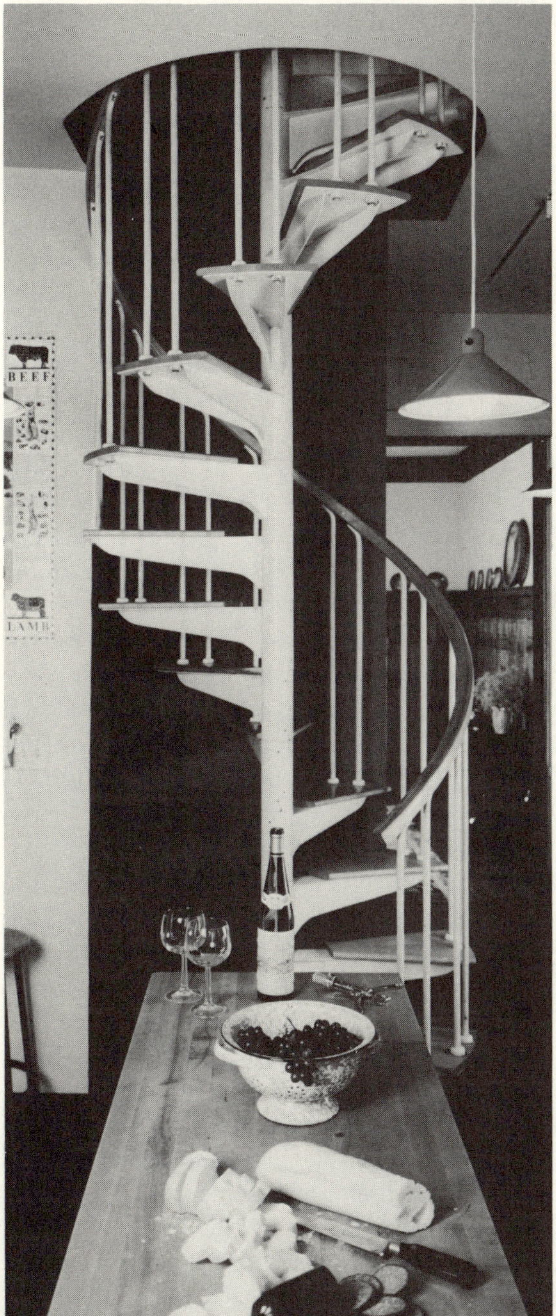

Installed in the kitchen as a second stair, this Duvinage sprial features a birch rail and wood-covered treads and platform. *(Duvinage)*

For anything wider than 5 feet or higher than 9½ feet, Duvinage recommends that the job be handled as a custom installation. But this doesn't rule out ordering a custom spiral and assembling it yourself. Duvinage preassembles all custom stairs in the shop and then takes them apart. Each part is numbered and carefully packaged. It may be one of a kind, but it's a kit nonetheless. We wouldn't put a Duvinage custom staircase beyond the competent d-i-y looking for a more elegant stair or a wider choice in materials, just so long as the parameters noted above are observed.

Duvinage Corporation
P.O. Box 828
Hagerstown, MD 21741-0828
Phone: 1-800-541-2645
Product literature: free brochure

FIREPLACES AND MANTELS

A working fireplace is high on the list of amenities of everyone's dream home. But traditional masonry fireplaces, the sort about which we tend to romanticize, have become prohibitively expensive. They are also woefully inefficient wood-burners and, as major air polluters, have been banned from new construction in many communities.

There's no need, however, to abandon the dream. Not when there are new factory-built fireplaces that can provide all the ambience of a conventional fireplace and at far less cost. The better factory-builts are designed to closely resemble a masonry fireplace in appearance and yet are five to ten times more efficient. Many of these new fireplaces are "zero-clearance" designs, which means they can be installed just about anywhere. Special footings are not required, and the insulated, double-wall design of the steel firebox allows for installation with "zero" space between the metal and standard wood-framing materials.

Modern zero-clearance fireplaces enable do-it-yourselfers to install a complete, high-efficiency fireplace system, much of which can be found in kit form. In a retrofit, zero-clearance fireplaces typically are installed against the wall rather than in the wall. Insulated double- and triple-wall metal chimneys come in sections that snap together to keep the fabricated flue system simple and economical. The stack is narrow and flexible enough that it usually can be run up through the roof without having to cut or displace any ceiling joists or roof rafters. Conventional construction materials are used to enclose the fireplace, which can be kept simple or, with the addition of a wood or marble surround and a mantel, can be trimmed to blend with any decor.

Just as with their antecedents, zero-clearance fireplaces have a firebrick core. Their efficiency comes through the use of tempered-glass doors, fans that draw room air through a heating chamber sealed off from the fire, warm it, and return it to the living area, and by bringing in outside air to support combustion. Almost no heated air is drawn from the room into the fire and up the chimney, and no vacuum is created within the home to draw cold outside air through cracks and crevices.

If you find dealing with a wood-burning fireplace too much of a bother, you can have a zero-clearance fireplace plumbed to burn natural gas, which costs about half as much for burn time as seasoned firewood and is twice as efficient. Behind glass, and with a gas-log insert molded from real logs, it's difficult today to tell a gas fire from a wood fire. Another attraction is that some gas models can be rigged to be operated with a hand-held remote control.

If you live in a condominium, an apartment, or other residence where the installation of a chimney is not possible, the old alternative was to install a dummy fireplace and an electric log insert or simply a mantel and firescreen for a purely decorative effect. But now there's a new generation of heat-circulating gas fireplaces that are chimney-free. These are direct-vent gas models, which bring in outside air for combustion and exhaust their fumes via a pipe within a pipe through the nearest outside wall. With no need for a chimney, a direct-vent gas fireplace can be installed beneath a window, in a bedroom, or even in the master bath. Airtight doors keep any noxious fumes from the home's interior. Unlike the characteristic blue flame of a conventional gas heater, these fireplaces burn with a yellow flame that is nearly indistinguishable from that produced by a wood fire.

You'll undoubtedly have to call in a plumber to make the gas connection. But even before you go shopping for a fireplace, whether gas- or wood-burning, check with the local building department to learn of any restrictions that may apply to the installation or use of fireplaces in the area where you reside.

Zero-clearance and direct-vent fireplaces can be found at fireplace specialty stores and many home and building supply centers. Most of these same sources also carry kits for wood mantels as well as surrounds in everything from precut marble and slate to false, but realistic-looking, brick and stone.

Basic zero-clearance, wood-burning factory-builts can be found for less than $500, but by the time you include the chimney kit, glass doors, an outside air kit, and construction materials, the entire package is likely to cost from $1,200 to $1,500. A realistic-looking gas-log set with burning embers can cost $400. You can add another $400 or more for a four- or five-piece marble hearth-and-surround kit, and from $350 to how-elegant-do-you-want-it? for a wood mantel kit.

A beautiful, symmetrically built fireplace is often the focal point of the home. Whether in a newly constructed dwelling or as a retrofit in an older one, a high-efficiency fireplace adds to the resale value of the home and can be one of the best investments the homeowner can make.

Superior Fireplace Company

When you hear the term "factory-built," the image that almost inevitably comes to mind is of a production line turning out an endless

Shown with a Superior Select Princeton mantel and Botticino marble surround, the HC-A fireplace comes in three screen sizes and with features specifically designed to boost warm-air output. *(Superior Fireplace Company)*

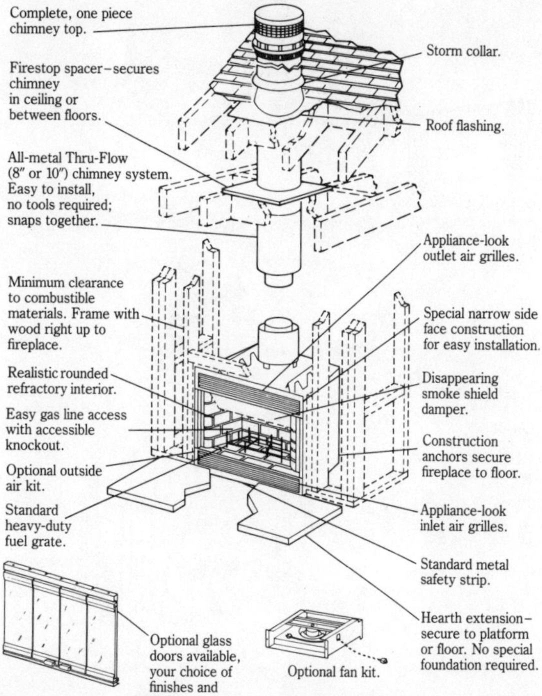

With Superior's zero-clearance design, installing the HC-A fireplace is surprisingly fast and easy. *(Superior Fireplace Company)*

stream of virtually identical products. Superior fireplaces may start out as factory-built fireboxes, but you would be hard challenged to find another company in this field that offers the end-design flexibility that this one does.

For starters, you would need more than a few production lines just to turn out fireboxes. Superior offers wood-burners and gas models in thirty-two styles and a wide range of sizes, from a zero-clearance gas-burner with a 27-inch opening to one of the biggest zero-clearance wood-burners in the industry, the TM-4500, which has a screen-opening width of 45 inches.

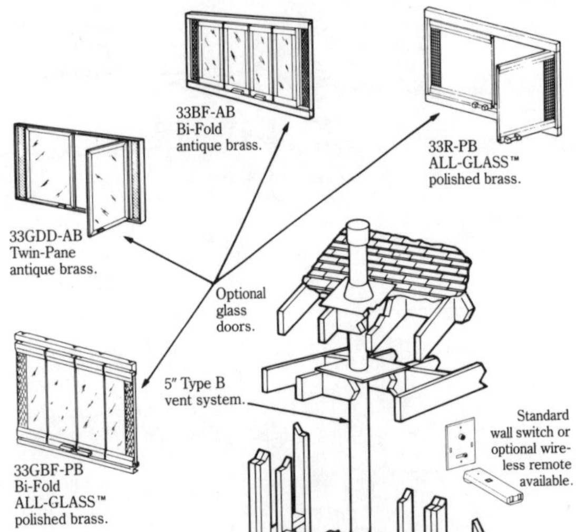

Shown is a typical home installation of the Superior GRD-5000, which offers a large 36-inch framed opening and is available in natural gas or propane models. *(Superior Fireplace Company)*

Nearly every Superior fireplace model offers a choice of decorator door and trim options, in stainless steel, polished brass, rustic antique brass, and contemporary jet-black. From the manufacturer's Superior Select line of premium fireplace accessories, there are genuine Italian marble surrounds and hearth extensions, sized to add classic elegance to any of its fireplace installations. To complete the overall presentation, Superior offers a variety of handsome hardwood mantels.

To return to that gas-burner with the 27-inch opening, it's Superior's G-436D Fireform, which calls for a 4-inch type B vent stack and installs easily into 16-inch-on-center framing. Offering all the coziness of a real open-flame fireplace, the Fireform, which is only 11 inches deep, makes a perfect second fireplace for the family room or den. Each Fireform comes standard with a 16,000-BTU natural gas burner, a rough-sawn-oak-design log set, and a pilotless electronic ignition system triggered by a standard wall switch. An optional wireless remote control kit is also available. An outside air kit is still another option.

You won't likely find a more realistic-looking, large-size factory-built fireplace than Superior's TM-4500. With its rugged refractory interior, massive screen opening, and stylish design, the TM (which stands for Traditional Masonry) easily provides the custom-built masonry look that many homeowners desire, plus all the cost-saving advantages of factory-built, zero-clearance construction.

If you are looking for the perfect solution for an installation where existing obstructions can't be moved, Superior's GPV-5000 Power Vent fireplace can be vented up, down, or sideways, over or around any obstruction, or through any wall, over a distance of up to 20 feet total with up to three 90-degree turns.

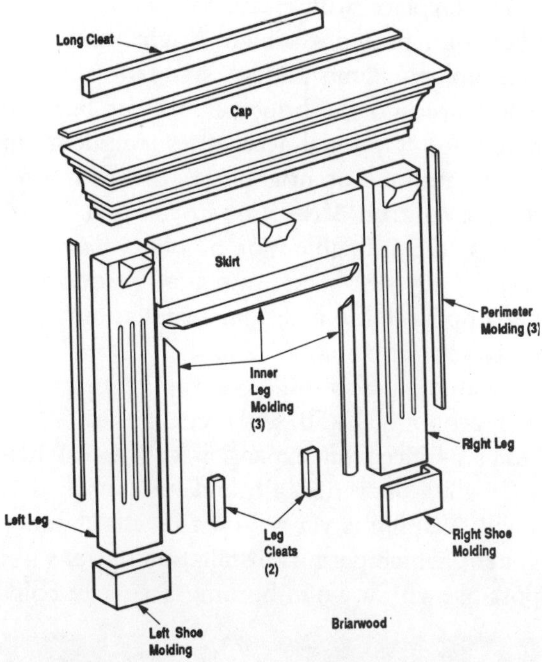

The Superior Select Princeton mantel, designed for use with Superior marble surround facings, can also be adapted to dress up other fireplaces. It comes in unfinished or prefinished oak. *(Superior Fireplace Company)*

The GPV's unique power-assisted 3-inch vent system permits installation where you want it, not where constraints dictate. It comes as a 24,000-BTU natural gas or a 21,000-BTU propane model, with or without an outside air combustion kit. The look of this heat-circulating unit is also hot and contemporary—a large glass enclosure accented by solid polished brass or black trim.

If you find it less romantic to sit before a wood fire that's behind glass, Superior also offers dual-role wood-burners that can be operated with the doors open or removed, or with the doors kept closed for greater heating efficiency. Direct-vent gas fireplaces that can go where no fireplace could go before are also included in this manufacturer's superior line.

Superior's precut marble-surround-and-hearth-extension kits, priced from $530, come in a choice of color selections: Carrara white, Botticino beige, and Bardiglio gray. The sets of polished Italian marble include a hearth extension, two surround legs, a surround header, and, for some installations, a surround filler piece. Sized to fit most Superior fireplace models, they can also be adapted to dress up many existing fireplaces.

The same would apply to Superior's mantel kits. They are designed for use with Superior marble surround facings, but can also accommodate many other surround materials. The mantels, priced from $370, come in paint grade, with a coat of white primer, or laminated hardwoods—red oak or poplar—to resist cupping, bowing, or splitting. The hardwood mantels come presanded, ready for finishing and easy installation.

Superior Fireplace Company
4325 Artesia Avenue
Fullerton, CA 92633
Phone: 714/521-7302

Heatilator

No major fireplace manufacturer is in a better position to capitalize on the trend toward efficient, heat-circulating gas-burners than Heatilator, which, while it has been building fireplaces for more than sixty years, has been producing gas models only since the late 1980s. Recent improvements in logs and flame technology, however, make Heatilator's newest gas-burners so realistic that few would ever guess that they aren't wood-burning fireplaces. With the push for cleaner air, this manufacturer fully expects that its gas-fireplace

Direct-vent gas fireplaces like Heatilator's heat-circulating GC300, which brings in outside air for combustion and exhausts its fumes through the nearest outside wall, can be installed beneath a window, in a bedroom, or even in the master bath. *(Heatilator, Inc.)*

line will account for at least half of its annual sales by the mid-1990s.

Heatilator's newest and most versatile gas appliance represents an improvement on conventional direct-vent gas models that must be placed against an outside wall. The GC300E can be vented vertically up to 6 feet and horizontally up to 11 feet, allowing the fireplace to be installed almost anywhere in the house. The sealed combustion system of this fireplace boosts its overall efficiency to 65 percent, while preserving indoor air quality. It draws no air from inside the home.

The fireplace, with Heatilator's new ceramic-fiber logs, provides a high, bright, and variable golden flame pattern simulating a real wood fire and the bright embers of burning hardwood. Optional accessories include a fan kit, an antique or bright brass trim kit, and remote control. Measuring 38½ × 41 × 24 inches, the 17-cubic-foot fireplace has been certified by the American Gas Association, has approval for bedroom installations, and costs approximately $1,265.

Heatilator also offers a top-venting, 27-inch economy ($450) gas fireplace, the G270. Add a 34-inch front expansion kit with cabinet-style glass doors ($200), and it's not all that small. Venting is via a 4-inch B-vent chimney system, which permits installation options not possible with a wood-burning fireplace. Fold-

The beauty of wood-burning, the convenience of gas, the option of remote control—all are provided by Heatilator's G270 Economy Gas Fireplace. The hardwood mantel and surround, also from Heatilator, come in ready-to-finish poplar or northern oak. *(Heatilator, Inc.)*

out nailing flanges and flat-bottom construction further simplify installation of the fireplace, which is only 11 inches deep. Figure another $300 for a set of natural-looking, hand-cast refractory logs. There's no stooping or bending to start the fire; a wall switch activates the unit. Or, if you find that too much bother, a hand-held remote control, and the necessary junction box for RC installation, is an option at around $120.

Fireside Furnishings gas-log inserts from Heatilator can be installed in most existing fully vented fireplaces with working flues. They operate off natural gas or propane and come in five different log lengths, ranging from 20 to 36 inches. At an operating cost of just pennies per hour, they can provide all the warmth and the golden flames of burning hardwood.

Fireside Furnishings also features a line of fine-quality hardwood mantels with optional wood surround. They come in poplar or northern oak and arrive ready to finish with your choice of paint or stain. The hardwood surrounds, adjustable to fit most 36-to-42-inch fireplaces, assemble easily, thanks to a quick-connect fastening system. They accept any ¾-inch facing material, such as marble, or can be installed flush on the wall, leaving room for decorative tile. Fireside Furnishings mantel sets are priced from around $510 to $620.

Not to neglect Heatilator's wood-burners, its newest heat-circulating fireplaces, the HB Series, are designed to give the most fireplace for the money without sacrificing Heatilator quality or reliability. These fireplaces can go almost anywhere, in any room, without masonry work or extra floor supports. Venting is via an 8-inch-inner-diameter/10½-inch-outer-diameter, two-wall, air-cooled chimney featuring snap-lock assembly. The chimney's small outside diameter allows installation through 16-inch-on-center joists without special firestops.

The fireplaces come with convenient nailing flanges for fast, secure installation. With their reduced clearance requirements to perpendicular walls, decorative surrounds, and mantels, the HBs are easy to frame and finish with conventional construction materials. Prices for the fireplaces start at $290 for a 36-inch standard model, the HB36A. Glass doors, an outside air kit, and an optional fan kit to boost the efficiency of the unit could easily bring the purchase price to $600 or more. A Heatilator chimney convenience pack, including three 48-inch chimney sections, roof flashing, and a terminal cap, would add another $290.

Heatilator, Inc.
1915 West Saunders Street
Mount Pleasant, IA 52641
Phone: 1-800-247-6798

Richards Studio

A fine mantelpiece not only catches the eye when you enter a room but makes the fireside a natural gathering place for family and friends. It expresses the heart of the home better than any other architectural statement.

Not every fireplace installation calls for a mantel, but where it is desirable to have one, Richards Studio (formerly Travis Millwork) provides an easy way to add an elegant handcrafted wood mantel to almost any fireplace.

Richard Studio specializes in building mantelpieces for designers, builders, and architects, high-end custom being about 75 percent of its output. But it also produces some of the finest mantel kits we've seen for assembly by the homeowner, and at prices that are easy to live with.

This authentic reproduction of a Federal-period mantel comes as an easy-to-build kit for $420. You simply trim the parts for a custom fit, then glue and nail the mantel together. *(Richards Studio)*

The ready-to-assemble kits, priced from $420, are prepared from hardwoods and composition ornamentation molded from antique carvings. Designed to achieve a sophisticated and tailored look, the mantels represent a weekend project for most d-i-y installers. They are sized to fit openings up to 42 inches high and 57½ inches wide. The mantel shelf itself is 74¼ inches long.

Some trimming of the skirt and the pilasters (legs) that make up the surround may be necessary to custom-fit the mantel to your fireplace. But there are detailed instructions to guide you here. The entire mantel can be assembled on the floor or workbench and then painted or stained before being installed. It could also be assembled directly on the wall—and then finished to match the room decor.

Richards Studio also offers custom mantels in classical, colonial, and English country style, priced from $775.

Richards Studio
Route 3, Box 848P
Spicewood, TX 78669
Phone: 512/264-2007

SAUNAS

The special, soothing benefits of the sauna bath have been known to the Finnish people for over two thousand years. But it wasn't until the mid-1960s that saunas began to be taken seriously in this country. Now there's a new generation of saunas, and they're becoming almost as essential as a good workout for those of us seeking relief from daily stress and an enhancement of health and fitness.

A sauna, which is Finnish for "bath" or "bathhouse," is taken in a tightly enclosed space that is lined with softwood and heated to between 160 and 190°F. while maintaining between 10- and 20-percent relative humidity. The dry heat, combined with steam produced when an occasional dipperful of water is poured over rocks brought to red heat on the grille of a wall-mounted electric heater, promotes perspiration. It's not only a natural way for the body to cleanse itself but it's uniquely revitalizing.

Saunas have also been revitalized. Instead of a wooden box to be set in a dark corner of the basement, they're now becoming design focal points and can go almost anywhere—bathroom, bedroom, family room, or fitness center. The newest trend is toward a greater use of glass, to enable the user to observe outside activities while reducing boredom and that closed-in feeling. Sauna manufacturers today are touching all bases, and if you want a sauna for the office, or for an apartment or condo where you can't do any major renovating, there are saunas that could work for you. There even are units that can be taken down and reassembled elsewhere, if the need arises.

Saunas are available in both modular and precut packages for easy assembly by the do-it-yourselfer. The smallest practical units, suitable for use by one person at a time, measure roughly 4 × 4 feet. But that's only the starting size. The biggest names in saunas can provide a preengineered unit to meet virtually any size or shape requirement. The one constant here is the height of the sauna room, which should not be more than 7 feet, to keep the heat down at bench levels.

No plumbing is required with a sauna, but a floor drain can be useful. It also helps if there's a shower nearby for washing and cooling-down purposes. The sauna bath routine calls for ten to fifteen minutes in the sauna, then stepping outside for a shower or cool-down before returning for another inning. However, no more than thirty minutes total should be spent in the sauna room during any one session.

A sauna room can be built in almost any level area, with concrete, ceramic tile, and one-piece vinyl the floorings of choice. The room should *not* be installed on carpeting or bare wood. Nor should paint or any other coating ever be applied to the interior surfaces of a sauna room or to the interior or exterior face of the sauna door. The exterior of the sauna room may be stained to suit a light or dark decor, however.

The most widely used sauna woods are Western red cedar and redwood. Even in the high heat of the sauna, these woods remain comfortable to sit and recline on. In a well-made sauna, benches are nailed from beneath, and there are no exposed nail heads or other metal fasteners that might burn the bather.

The most important component of any sauna is the heater. It must be carefully matched to the room design. If the room is not sufficiently well-insulated, or if the design uses more glass than another design of equal size, a heater of larger capacity will likely be needed. But the manufacturer or his representative can help you here.

As a general caution, pregnant women or persons with a heart condition or in poor health should consult their physician before using a sauna.

Am-Finn Saunas

An industry leader since 1962, Am-Finn offers both precut and easy-to-assemble prefab cedar sauna kits. This company's freestanding modular sauna rooms are of particular interest and less expensive than most. Complete portability is a major plus.

With its prefabs, Am-Finn uses a system of panels—a 2½-inch sandwich of 1 × 6 tongue-and-groove (T&G) Western red cedar, insulation, and brown oak veneer plywood and cedar trim—that are shipped ready for quick assembly. For a typical 6 × 4-foot Am-Finn modular sauna room, there are just nine wall sections and two ceiling panels. Assembly takes approximately 4 hours. It can take even less time to disassemble the unit for a move to another location.

The standard package includes a prehung cedar door built especially for sauna use, preassembled heavy-duty cedar bench tops, vaporproof light fixture, heater guard fence, duckboard, and a commercial-grade, all-stainless-steel electric heater, complete with thermostatic controls and high-density sauna rocks. The standard door includes a 9 × 18-inch window, but much larger insulated glass panels are offered, and they are an increasingly popular option with Am-Finn customers.

Large-scale insulated vision panels that enable sauna users to observe outside activities are becoming a popular option. *(Am-Finn Sauna Co.)*

Am-Finn's sauna planning kit includes an audiotape and general instructions for both precut and modular sauna installations. *(Am-Finn Sauna Co.)*

Sauna rooms should be on a separate, or dedicated, electric circuit, and Am-Finn recommends that a licensed electrician be called in to make the final electrical connection.

The prefab package for a 4 × 6 × 7-foot Am-Finn modular sauna room with a 4.5KW heater costs $3,050. Other sizes available range from 4 × 4 to 10 × 10 feet and are priced from $2,375 to $6,800, with an appropriately sized heater.

Am-Finn Sauna Company
P.O. Box 810
Valley Forge, PA 19481
Phone: 1-800-237-2862
Product literature: free color brochure

Amerec Saunas

The fact that Amerec saunas have been installed everywhere from the White House to NASA headquarters and the Chicago Bears Training Center is a reflection not only of the quality of this manufacturer's product but of the ever-widening recognition of the sauna's many health benefits.

With thirty-eight standard precut sauna room sizes, plus twelve standard modular rooms and a personal sauna, glass-front options and four styles of sauna heaters, as well as the ability to create custom saunas in any size or shape, Amerec is in a unique position to meet the special sauna needs of health-conscious residential customers and commercial interests alike.

Amerec's precut sauna rooms, fully half of which are ordered in custom sizes, are its most popular package for both residential and commercial applications. Keep in mind, however, that these are built-ins. Each precut-room package contains all the material and equipment needed to complete the sauna room *after* the room has been framed, wired, and insulated, and wallboard has been applied. Even the finish work of the exterior walls is left to the customer or builder.

All Amerec precut packages include 1×4 T&G Western red cedar for walls and ceiling; interior corner, base, and ceiling trim; cedar benches and supports; heater guard rail; door; sauna heater and controls; duckboard floor in walking area; bucket; dipper; light; thermometer; head/back rests; and sauna rocks. Rooms range in size from a one-person 3 × 4-foot room to a large, usually commercial, 12 × 12-foot room. Prices, excluding heater and controls, range from $1,669 to $5,572.

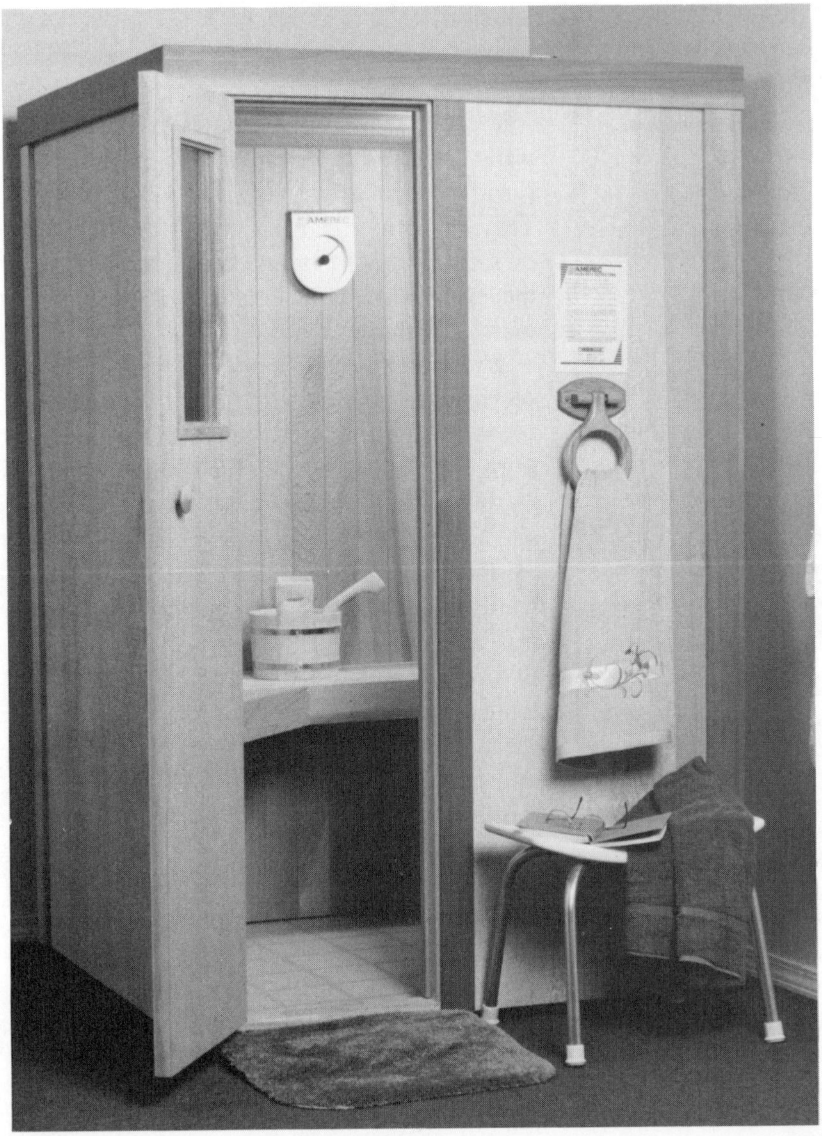

Amerec's prefab Personal Sauna ($2,695) is ideal for apartment and condo dwellers who are barred from doing major renovations. *(Amerec Products)*

The heater and controls are excluded from the room prices to give greater flexibility in choosing a different KW size or style heater. In addition to the traditional Finnish wall heater, Amerec offers a triangular heater that nestles into the corner of smaller sauna rooms, a floor heater to accommodate larger, normally commercial-size sauna rooms, and a wall heater with electronic controls.

The electronic sauna heater combines the time-tested qualities of the Finnish heater with advanced computer technology. With the 7-day programmable control mounted at a convenient location outside the sauna room (it can be up to 50 feet away), it takes only a touch of a button to have the room heated and ready for use at a specific time of day.

An electronic wall heater suitable for a 32-

Amerec's precuts are designed for permanent installation in new construction or where remodeling doesn't present a problem. *(Amerec Products)*

Vertical or horizontal application can be specified for the Western red cedar walls of an Amerec precut sauna room interior. *(Amerec Products)*

square-foot room of standard 7-foot height would cost $885, plus another $227 for the electronic controls. A triangular sauna heater (with built-in thermostat-timer controls) for a room of this size would cost $725.

If you're not up to the building skills called for with a code-complying, permanent installation, Amerec's freestanding modular sauna rooms can make the job a lot easier. These rooms are easily assembled from prefabricated wall and ceiling panels with an interior facing of 1×4 V-joint Western red cedar, fiberglass batt insulation (rated R-11) in the core, and a smooth mahogany veneer exterior. Cedar on the outside face is offered as an option. The outlet box for the light fixture and the junction box for the heater connection are preinstalled and tied together with conduit extending through the ceiling for easy hook-up.

Amerec's modular saunas are designed to be completely portable and can be taken down and reassembled in a new location with ease. Full-length glass panels, interchangeable with any standard 22½- or 24-inch wood front panel, are available as an option. Double glazing serves as insulation against heat loss, but it is recommended that, for every two glass panels ordered, the sauna heater be upgraded.

Prices for Amerec modular prefab sauna rooms range from $2,596 for a 4 × 4-foot room to $6,212 for a room measuring 8 × 12 feet. Again, sauna heaters and controls are offered separately. Otherwise, the package includes the same accessories mentioned above with the precuts.

Amerec's Personal Sauna is ideal for apartment and condo dwellers who are barred from doing major renovations. The unit em-

Prefabricated wall panels and bench tops make it easy for one person to assemble an Amerec modular sauna room in a few hours. *(Amerec Products)*

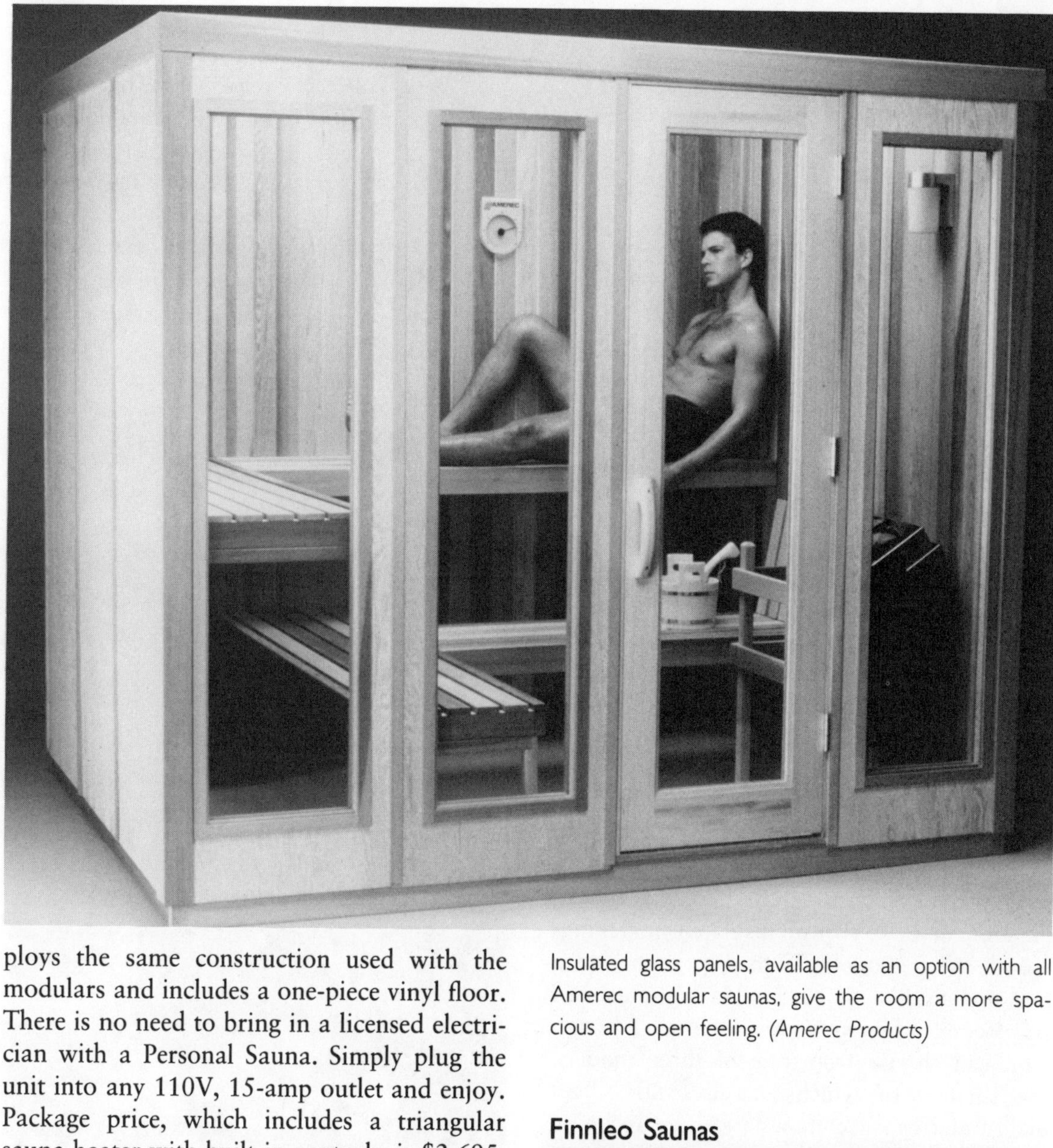

Insulated glass panels, available as an option with all Amerec modular saunas, give the room a more spacious and open feeling. *(Amerec Products)*

ploys the same construction used with the modulars and includes a one-piece vinyl floor. There is no need to bring in a licensed electrician with a Personal Sauna. Simply plug the unit into any 110V, 15-amp outlet and enjoy. Package price, which includes a triangular sauna heater with built-in controls, is $2,695. The unit measures 48 × 42 × 74 inches.

Amerec Products
P.O. Box 40569
Bellevue, WA 98004
Phone: 1-800-331-0349

Finnleo Saunas

Finnleo is the only company we know to offer a traditional Finnish log sauna in precut kit form. Built up on a base frame and reinforced with vertical tie rods, precision-milled 2¼ × 4½-inch tongue-and-grooved timbers are sim-

Precision-milled Nordic white spruce timbers are fitted together like Lincoln Logs to form the walls of a Joutsen sauna room. Full-width headers complete the walls and the opening for the door. Overlapping notched corners ensure a rigid, precise fit. *(Finnleo Saunas)*

ply stacked to form the sauna-room walls. Overlapping notched corners ensure an exact fit. A periodic tightening of the threaded tie rods guards against air leaks and any warping of the timbers. Since the solid timbers provide insulation as well as interior and exterior finished wall surfaces, the system could also be said to be modular.

The roof system for these saunas *is* modular and comes as two preinsulated, matching-wood ceiling panels. With a prehung door and prebuilt benches, a Finnleo log sauna can be assembled by two people in under 2 hours. You can choose from one of three models, the smallest of which measures 80 × 80 inches and can accommodate three reclining or six sitting. There are five door options, too. Prices for these saunas, known as the Joutsen series, range from $3,900 to $6,200, including an appropriately sized heater, controls, and accessories.

Finnleo saunas are built with fine-grained Nordic white spruce instead of the darker

The ceiling for the sauna room comes as two preinsulated, interlocking panels designed to match the V-grooved log walls. *(Finnleo Saunas)*

cedar and redwood used by most other major sauna manufacturers. As a further distinction, benches, backrests, trim, heater guards, and duckboards are made from African white abachi, an exotic wood that has no knots or splinters and remains comfortable to the touch even at the highest sauna temperatures.

Finnleo also offers conventional panelized saunas that are nearly as easy to assemble as its Joutsens. Its popular Sisu saunas range from a 45-inch-square Mini Sisu, the perfect sauna for two, to a five-sided unit with a floor area of approximately 40 square feet. If you're looking for contemporary European styling, Finnleo's luxuriously detailed Designer series offers everything from entrance atriums and dramatic window treatments to recessed lighting and six-sided designs.

With more than fifty years of sauna experience, Finnleo has built an enviable reputation for its complete, ready-to-install packages that two can assemble in a matter of hours. Prices for its prefab saunas range from $1,995 to $13,995.

The wall-mounted Finnmatic heater for the 80 × 80-inch sauna comes with built-in controls and 57 pounds of peridotite sauna stones. *(Finnleo Saunas)*

Finnleo Saunas
One West Lakeshore Drive, P.O. Box 610
Cokato, MN 55321
Phone: 1-800-FINNLEO

Outdoor Structures

GAZEBOS

The gracious appeal of the gazebo, which for thousands of years has been the centerpiece of elegant garden settings, still endures. As a reminder of a quieter, simpler time, many today are turning again to these enchanting structures for the tranquility they can, if only for a little while, bring back into their lives.

A gazebo not only can provide protection from the sun or a sudden shower, it's a place to escape to for a quiet read, or simply a perch from which to enjoy the view. It's a magic place where daydreaming is accepted and children can play during the day. It can also be an enjoyable setting for a quiet luncheon or a gala wedding.

The typical gazebo is octagonal, though there are also gazebos that are square or rectangular in design. Styling is mostly Victorian, with Oriental overtones. Wooden scrollwork, or gingerbread, is a hallmark of the classic Victorian gazebo. The Oriental influence is mostly reflected in curved, pagodalike roof designs. Tradition dictates that it have an illusion of delicacy, and yet possess the hidden strength to withstand the tempests of time.

Local building codes and lot-line-setback requirements should be checked out before you make a final decision on the placement of a gazebo. You'll also want to take wind, shade, and sun patterns into consideration. By all means, take advantage of a view. But it's equally important that the structure, which will become a focal point, be situated to complement its surroundings.

Gazebos may be erected on any level surface, whether a stretch of lawn, a wooden deck, or a concrete slab. In a garden setting, most are set atop poured-concrete pilings or pressure-treated 6×6 or 8×8 posts set in the ground below the frostline. Kits typically come in easily assembled sections, predrilled and ready to be bolted together. At most, with the pilings in place, it shouldn't take more than a day for two or three people to put up a gazebo of less than bandstand size.

Decay-resistant cedar or redwood are the woods most commonly used in gazebo construction. Most homeowners choose to leave the wood surfaces natural. Applying paint or stain, however, can give any of these structures an entirely different character, and may better complement a garden or other setting. Painting is strongly recommended if the gazebo is constructed with poplar, fir, or any of the other softwoods. With these woods, we would suggest two coats of an oil-base primer followed with two top coats of an oil-base finish paint.

Gazebos are often topped off with a cupola or weather vane. A ceiling light is another finishing touch. Most sources of gazebo kits can also supply appropriately designed tables and benches, as well as a full set of screens for use where mosquitoes and other flying insects can be a problem. Gazebo kits, in cedar, start at under $2,000 for octagonal units with a diameter of 8 or 9 feet.

Dalton Pavilions

There are no nuts and bolts or other structural fasteners in plain view in a Dalton gazebo. With nearly forty years of experience in structural engineering and design of outdoor pavilions, this family-owned-and-operated business goes the extra step and caters to those who would find any indication of a prefabricated structure unacceptable.

Dalton produces octagonal gazebos in seven sizes, ranging from 9 to 26 feet in diameter. Crafted in the finest woodworking tradition from clear Western red cedar, which has its own built-in preservative and needs no treatment, major components of every Dalton gazebo are preassembled using stainless-steel screws and other corrosion-resistant fasteners. Floor sections are blind-nailed to leave no surface evidence of nailing. All models include a switch box and a concealed electrical channel for wiring by the homeowner.

In most states east of the Mississippi, arrangements can be made to have a Dalton gazebo delivered and installed by a Dalton crew. But ordered as a kit, prefabbed in large sections and predrilled for final on-site assembly, two people with little carpentry experience should be able to complete an installation in from 8 to 16 hours. Most of the components are joined by bolts, brackets, and threaded rods. Trim strips and other construction techniques cover any screw heads or other fasteners that would otherwise be visible.

Gazebos like Dalton's elegant Williamsburg Pagoda model, which incorporates elements of both colonial and Oriental architecture, can add a pleasing dimension to virtually any outdoor setting. *(Dalton Pavilions)*

The timeless design of Dalton's Classic series gazebos blends with many styles of architecture. The seven-model series includes three Classic Pagodas, offers sizes ranging from 9 to 15 feet. *(Dalton Pavilions)*

There are three basic Dalton gazebo designs: the Classic, a conservative design with clean, elegant lines; the Williamsburg, with the legacy of colonial architecture; and the Victorian. All three designs can also be ordered with a tiered pagoda roof. The smallest pagoda-styled Dalton gazebos are 11-footers, and they are priced from $6,573, for the Classic Pagoda. The least expensive of the gazebos is the 9-foot Classic, priced from $4,606. You can shave about $700 from this price if you order the gazebo without the floor. Concealed hardware is provided for anchoring the gazebo to a patio or other level, secure base. Dalton's top-priced model is the 26-foot Williamsburg Pagoda, at $15,570 (or $12,450 without the floor).

Optional screening is available for all models. The screening package includes underfloor screening, removable dark side screens with bronze-colored frames, and a matching screen door. Screening packages are priced from $935. Other options include octagonal tables, in clear Western red cedar, priced from $385 for a 40-inch size; sets of cedar benches, from $246 for four; a cupola with finial, at $250; and a weather vane, choice of eagle or horse, $100.

Dalton gazebos are manufactured to precise dimensions and prefabbed in large sections to assure easy on-site assembly. *(Dalton Pavilions)*

Dalton Pavilions, Inc.
7260–68 Oakley Street
Philadelphia, PA 19111
Phone: 215/342-9804
Product literature: color catalog, $3

Handy Home Products

Sold through Sears, Scotty's, and other home centers, the Handy Home Products line of ready-to-assemble outdoor structures includes low-priced, all-cedar gazebos in five different configurations. In addition to the more familiar octagon, there are ten- and twelve-sided models, as well as square and rectangular designs. The line even includes a gazebo-styled sandbox.

Like all Handy Home kits, the gazebos are preengineered, with components precut and major wall and roof sections preassembled at the factory, making the on-site installation (with a Phillips screwdriver) easy for the average do-it-yourselfer. Estimated assembly time for the 9-foot-9-inch octagon is only 2 to 3 hours. The larger sizes take a bit longer. The

Sold through Sears and other home centers, Handy Home's model G-99 cedar gazebo, with the optional floor kit, screens, cupola, and weather vane, could be installed complete for under $2,650. *(Marco Wood Products)*

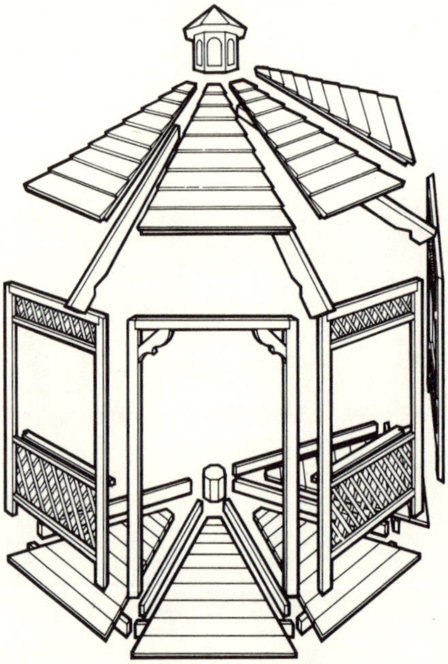

With major floor, wall, and roof sections preassembled at the factory, Handy Home Products makes on-site installation of its gazebos easy for the do-it-yourselfer. *(Marco Wood Products)*

red cedar will turn a silvery gray as it ages, but if some other color is preferred, a quality paint or stain can be used with good results on cedar.

Base prices start at $1,300, for the 9-foot-9-inch octagon, and go to $2,500 for the 14-foot-5-inch dodecagon and the 11-foot-3-inch square. The last gazebo comes with a skylight. Options include modular cedar floor kits, priced from $400. Floor kits can take less than half an hour to assemble. A cedar cupola, from $120 to $200, depending on size, is another option. The cupolas, which have carved designs on each face for a custom touch, come preassembled and predrilled for easy installation. Weather vanes, priced from $50 to $80 in a choice of three designs, are also offered.

Handy Home's S-1111 square gazebo measures 11¼ feet on a side and comes with a skylight. Kit package price, including optional modular floor and removable screen panels and door, is $4,200. *(Marco Wood Products)*

Screening is another popular option with these outdoor havens. A full set of Handy Home's cedar-framed door and wall panels with bronze-colored fiberglass screening costs from $700 to $1,000. The removable panels are predrilled for easy attachment to the structural framework of the gazebo.

Handy Home's gazebo-styled cedar sandbox is a six-sided design featuring a latticework roof topped with a finial. The 73-inch-diameter box is spacious enough to accommodate at least six children seated along its rim and provides enough sand area for many hours of playtime fun. Kit price of the Golden Carrousel sandbox is $250.

Marco Wood Products
P.O. Box 548
Walled Lake, MI 48390-3222
Phone: 1-800-221-1849
Product literature: free Handy Home Products catalog

BowBends

We don't know of any other company that comes close to the creativity exhibited in BowBends' pool and garden structures. A division of Bow House, which is renowned for its unique Architectural Package of reproduction building materials (see page 141), this Massachusetts manufacturer offers gazebos, arbors, ornamental bridges, pool houses, and other structures that, while ordinary in function, are a delight to the eye.

BowBends gazebos, with which we are primarily concerned here, include the Shandy, an economy model with an asphalt-shingled roof; the Tea House/Pagoda, with copper fishscale shingles; and the classic Belvedeary, which is constructed like a house and, if maintained,

BowBends' curved-roof Belvedearys can be ordered with a variety of closures, including glass, for an all-weather structure that can be heated or air-conditioned. *(BowBends)*

The Shandy is a BowBends economy-model gazebo. Offered in diameters of 10, 12, 14½ and 17 feet, it comes with tan asphalt shingles and is priced from $2,605. *(BowBends)*

should last as long. They all can make a lovely focal point and are offered in kit form for nonprofessional assembly.

By definition, a gazebo and a belvedere are virtually synonymous, but a belvedere more specifically is a small summer house built for giving a view of the scenery. BowBends' curved-roof Belvedearys, which are offered with a variety of closures, including open rail, screen, louver, lattice, and glass panel, are more house than other manufacturers' gazebos. They are all-weather structures, and glass-enclosed versions can be heated or air-conditioned.

On-site construction of a Belvedeary presents more of a challenge than would a typical bolt-together gazebo kit. BowBends, for example, recommends "invisible caps" on the white cedar roof, which is time-consuming but classic. More attention is also paid to interior details, to make the structure as attractive on the inside as on the outside. A Belvedeary

A bathhouse based on a seventeenth-century privy exemplifies how BowBends can take a structure that is ordinary in function and make it a delight to the eye. *(BowBends)*

Ornamental bridges are another BowBends specialty. This is the Chippendale, one of the manufacturer's six bridge designs. Price of this 16-foot span is $3,435. *(BowBends)*

with the invisible caps can easily take 80 to 90 man-hours to erect.

BowBends offers the Belvedeary and its other gazebos in 10-, 12-, 14½-, and 17-foot (sometimes referred to as pavilions) diameters. Open-rail Belvedearys are priced from $4,845. The Tea House/Pagoda, with copper shingles, is priced from $5,605. The Shandy, which is a simpler structure with a straight, lower roof pitch, is priced from $2,605 with tan shingles. Octagonal deck kits, priced from $760, are available for all models.

Mixing and matching is possible. BowBends has done Shandys with curved roofs and Belvedearys with straight roofs. Nor is there any reason that all eight sides have to be the same. A BowBends gazebo can be used to focus a view.

Ornamental bridges are another BowBends specialty. There are six standard designs, including colonial and Oriental motifs. Bridges can be any length up to 80 feet, any width up to 8 feet, and any radius. They are constructed from pressure-treated southern yellow pine. Prices for a typical BowBends 8-foot span start at around $1,600; for a 16-footer, the range is from $2,945 to $3,435.

BowBends of Bow House, Inc.
92 Randall Road, P.O. Box 900
Bolton, MA 01740
Phone: 508/779-6464
Product literature: brochure, $3

GREENHOUSES

Greenhouses are for growing. They do not make very good solariums. Nor do solariums or other sun rooms make very good greenhouses. With a greenhouse, you control the climate and create the garden world of your choice, whether that's for exotic, heat-loving tropicals or for fresh vegetables all winter long. Increasingly, a greenhouse is the first entry on the wish list of true garden lovers.

The traditional greenhouse is a freestanding wood- or aluminum-framed glass structure admitting light through two sides, gable ends, and a span roof. Some are little more than stark glass boxes. But others are designed to blend with the architecture of the home, without any sacrifice of function or workspace. Among the design options: bronze-tone or white enameled aluminum framing, graceful curved eaves, slanted side walls, masonry basewalls.

It's architectural details such as these that have made lean-to greenhouses, designed to be installed against an outside wall of the house, among the most popular of today's greenhouses. A lean-to, figuratively, is half of a freestanding greenhouse that has been split down the middle. Arrange the installation to cover a patio door or a picture window, and you will have a panoramic view of your garden-under-glass at any time of day or year, regardless of the weather.

However, for serious plant growing, you need plenty of morning sun as well as overhead sunlight. Thus, it's the role of old Sol, more than anything else, that will determine the location of the greenhouse, and ultimately its shape, whether lean-to or free standing. Most commercial growers favor a slightly NE to SW orientation of the greenhouse. A tree that provides afternoon shade through the summer months and loses its foliage in the fall can also be an asset.

A major manufacturer of greenhouses may offer a hundred or more standard models, but what this usually comes down to is half a dozen series, each of which may include both lean-tos and freestanding greenhouses in three or four different widths, and in a choice of straight- or curved-eave models, for either glass-to-ground or built-up (calling for a knee or basewall) installation. Lengths typically start with three or four sections and may go to twenty-five or more.

Manufacturers caution against ordering a greenhouse that could later prove too small. A 10-foot width will easily accommodate a 3-foot bench along each side wall. For three rows of benches and two aisles, you'd need a 14-footer. Double staging tables could accommodate additional plants or seed flats where you have glass-to-ground side walls. Sufficient interior height to accommodate hanging plants or allow for the installation of glass shelves should be another consideration.

We are of the school that holds that a residential greenhouse should present as pretty a picture when viewed from a distance—the house, say—as when you are enjoying the quiet beauty inside the structure. Cloudy fiberglass and some of the other plastics that are among the greenhouse glazing options offered today at best detract from this view. While we might be willing to accept fiberglass roof coverings, where breakage from hail could be a problem, or as a filter to reduce summer heat build-up, we would still want to keep to glass walls. Glass not only transmits the most light but *looks* better in the backyard or garden. Unlike some of the plastics, glass doesn't discolor, fog, or fade over time.

It's also easier to keep clean. The glazing of choice in greenhouses is double-strength, or DSB, horticultural-grade clear float glass.

Adequate ventilation is crucial to the success of any greenhouse, not only to avoid overheating but to provide the air exchange necessary for healthy plant growth. Roof venting along the ridge line is highly desirable. Other systems incorporate operable windows or a louvered opening in a gable end or side wall, with an exhaust fan mounted in an opposite wall. The ideal is full-length roof vents combined with cross ventilation. Regardless of the venting system used, it should be automatic in operation. Automatic single-vent openers can be purchased for as little as $37. Controllers for full-length ridge venting can cost $285 and more, but then so can an 18-inch shutter-mounted exhaust fan.

A tip offered by Texas Greenhouse Company for a completely automatic ventilating system is to install a shutter-mounted fan in one end of the greenhouse and a motorized shutter in the opposite end. Both apparatuses can be controlled with a single thermostat. Such a system could be put together for between $300 and $500, depending on the size of the fan needed for the area to be vented.

Heating is another consideration. To control the seasons for year-round growing in most sections of the country, a heater of some type, sized to meet the cubic footage of the greenhouse and the inside and outside temperatures that will be involved, is an inevitable need. Through the colder months, the sun will supply most of the heat needed during the day, but supplemental heat will be required at night. Electrical space heaters are the most popular of the unit heaters, but gas and oil heaters are also available. For a modest-size greenhouse, you're again looking at an item in the $300 to $500 range. With a greenhouse that is close to or attached to the house, you might want to investigate the possibility of hooking into the home-heating system as an alternative to a separate heating plant.

The folks at Sturdi-built Greenhouses report that many of their customers, in really cold winter climates, conserve on fuel by lining the inside of the greenhouse with clear polyethylene plastic or clear vinyl, just for the winter months—which introduces the question, why isn't insulating glass recommended for greenhouse glazing? For one, there's the cost factor. It's an extremely expensive alternative. We are also advised that, with factory-insulated glass, the higher humidity levels found in greenhouses can eventually cause fogging and condensation between the sealed glass layers, resulting in deterioration.

Cooling the greenhouse during exceptionally hot weather can be as important as keeping the temperature up during those icy winter nights. In the deep South, you might want to consider the purchase of an evaporate cooler, which can lower the inside temperature by 20 to 30°F. and circulate the cool humid air, on which orchids and most other plants thrive, throughout the greenhouse. But a much less expensive alternative is to counter the effects of high temperature with shade and induced humidity along with proper ventilation.

There are fiberglass shading panels, as mentioned earlier, but economical polypropylene and plastic shade cloths are also used extensively. These are offered in varying weaves to provide the percentage of shading needed. Roll-up aluminum shades are considerably more expensive but can easily be adjusted from outside as weather and sunlight vary. They are aesthetically pleasing and can provide all-weather protection and insulation as well.

A glass greenhouse is a weighty item and requires a solid footing, the depth of which will likely be determined by local building codes. The local building department could also have some say in what materials are used in a foundation wall. As for the floor of the greenhouse, some choose to go with brick or ceramic tile and a set-in drain. Aggregate or dirt floors with a brick path are more common, though; they never need mopping and can even be watered down to help provide the humidity needed for the growing of most plants.

Janco Greenhouses

One of the largest residential greenhouse manufacturing companies in the United States, Janco offers a standard line of aluminum-and-glass structures that includes both freestanding and lean-to models, curved-eave or straight-eave, built-up (to sit on a 30-inch-high masonry wall) or glass-to-ground (requiring only a foundation to 8 inches above grade). In all, there are more than one hundred models, sized to fit the smallest lot or the largest estate.

The typical Janco greenhouse is not difficult to erect. Each precut extruded-aluminum part and lite of glass is marked and identified to correspond with a set of plans for the particular model. Detailed instructions for do-it-yourself assembly are provided, but, should professional help be needed, the local Janco dealer can recommend a contractor.

Installation of a Janco greenhouse calls for a continuous concrete footing to a minimum depth of 20 inches (local building codes may call for more). If requested, foundation plans will be provided in advance of the greenhouse delivery. With the perimeter foundation wall, the greenhouse can be made to fit almost any terrain.

Full-length roof vents are standard on nearly all Janco greenhouses. The exceptions would be a few of the smaller, lower-profile lean-tos. Tempered glass, which has four to five times the strength of annealed glass, is standard on many of Janco's lean-tos, making them acceptable under the most stringent building codes applicable when the home itself will be part of the structure. Glazing for all other models is double-strength greenhouse quality in the standard 24-inch width, which means that replacement glass should be more easily

Janco's curved-eave Camellia 14 includes full-length venting on both sides of the ridge and is priced from $4,001 for a 5-section, 11-foot built-up. For each extra section, add $416. *(Janco Greenhouses)*

The glass-to-ground Camellia 14 costs $5,698 for a 5-section, 11-footer (as shown), and accommodates three 36-inch-wide Janco redwood benches. Add $601 for each extra greenhouse section. *(Janco Greenhouses)*

obtainable. Factory-sealed insulated glass is offered as an option.

Aluminum glazing bars for curved-eave models are one-piece, from ridge to sill, for sturdier construction. In all Jancos, the glazing bars are so designed that glass shelves and hanging baskets can be installed without any drilling of the framing members. The maintenance-free aluminum comes with the standard mill finish, but a bronze-tone enameled finish is available as an option.

Janco lean-to greenhouses are priced from $2,123, for the 5½ × 11-foot, curved-eave, built-up Rancher 6. The glass-to-ground Rancher 6 costs $300 more. Prices for Janco freestanding greenhouses start at $3,261, for the 10½ × 11-foot, curved-eave, built-up Senator. For such options as a bronze-tone finish, add $418; for two redwood benches, $634; for Janco Aut-A-Matic ventilation (which employs a motorized intake louver, an exhaust fan, and a thermostat), $195.

Janco greenhouses and accessories are sold through a nationwide network of independent Janco dealers.

J. A. Nearing Company, Inc.
9390 Davis Avenue
Laurel, MD 20723-1993
Phone: 1-800-323-6933
Product literature: catalog, $5

Spacious, straight-eave glass-to-ground Estate models are priced from $3,964. With nine sections, as shown, price is $5,548. Aluminum shades and the Dynavent heater in the gable are extras. *(Janco Greenhouses)*

Sturdi-built Greenhouses

Although sold at factory-direct prices, Sturdi-built greenhouses are not mass-produced. They are designed and crafted to meet the special needs of each customer, whether in Alaska or Florida. This is a small, family business, dedicated to providing personalized service before, during, and after the sale of its products.

Sturdi-built greenhouses are constructed with redwood and glass, and come either as precut or prefabricated packages. The prefabbed models are shipped in large, factory-assembled sections, and with a prehung door, for quick-and-easy, do-it-yourself installation. With precut models, for those who wish to save by doing more of the assembly themselves, only the endwalls are preassembled, and the door is not prehung.

All units are shipped unglazed. Precut glass is furnished and must be installed after the greenhouse framing is erected. Specially milled redwood trim caps that fit onto the glazing bars after glass is installed are a standard feature of Sturdi-built greenhouses. They add another dimension to the fine appearance of these structures and serve to protect and extend the service life of the glazing system. Basewalls, to a height of 32 inches, are a feature of a number of Sturdi-built designs. These, in redwood, can be ordered as part of the package, complete with vents for natural air flow, or they can be built on-site with locally obtained materials.

The manufacturer offers greenhouses in both freestanding and lean-to models, with vertical or slanting walls. Roof vents that can be easily automated are built into each design. Glass is the standard glazing, but fiberglass may also be incorporated into the design for hail-proof construction. Doors are handcrafted of 1½-inch-thick redwood. Drip-collection grooves in rafters, together with interior gut-

Prefabricated redwood-and-glass Tropic series greenhouses come in both freestanding and lean-to versions and are a favorite with Sturdi-built customers, particularly in colder climates. *(Sturdi-built Greenhouse)*

Sturdi-built prefabricated greenhouses come in large, factory-assembled sections for easy do-it-yourself installation. Precut glazing isn't added until after the framing has been erected. *(Sturdi-built Greenhouse)*

ters and downspouts, prevent condensation from ruining plant foliage and blooms or dripping onto the greenhouse worker.

Options include jalousie windows that can be mounted in an endwall to provide extra ventilation. These come complete with screens and "storm windows." There's also Sturdi-built's unique Thermal Option, or double-glazing system, which insulates and shades the greenhouse. The system will accommodate a variety of easily removable glazings, which solves the usual problems of double-glazed greenhouses. The Thermal Option can be added to all of the greenhouse or any part of it. For example, fiberglass could be added to the roof for both shading and insulation.

The attractive glass-to-ground design of the 11 × 14-foot Solite provides extra growing space under the benches. Redwood basewall with built-in vents is included in the pre-fab's $4,007 price. *(Sturdi-built Greenhouse)*

Walls could be done later as desired. The Thermal Option can be added to any Sturdi-built greenhouse at any time. All framing members are milled to receive it.

Prices for Sturdi-built's precut models start at $1,888 for the freestanding 8 × 9-foot glass-to-ground, slant-wall Solite Junior. A 5½-foot-wide Solite Junior lean-to in the same 9-foot length costs $1,578. Lengths up to 14 feet are available. A Dutch door is an option at $45.

Prices for prefabricated models start at $2,658 for the freestanding 9 × 12-foot Tropic slant-wall, built with a 32-inch basewall. Add $546 for the basewall package. For a Dutch door, add $65. The Thermal Option would cost $879. The Tropic is also offered in a freestanding 10-foot width as well as in 7- and 9-foot lean-to versions, at lengths of from 12 to 20 feet.

The Sturdi-built line also includes the Tropic-Twin, which combines a utility/potting room with a Tropic greenhouse. Special self-shading fiberglass covers the utility area. The prefabbed Tropic-Twin, which measures 10 × 20 feet overall, costs $5,986 with the 32-inch basewall package.

Sturdi-built also offers the prefabricated Flower Gazebo, a compact, 8-foot-diameter, glass-walled octagon that is perhaps more showcase than working greenhouse. It's priced at $2,887. There's also the circular Sunflare, in 10-, 12-, and 14-foot diameters, priced from $4,667.

This manufacturer absorbs the freight costs by common carrier on all of its greenhouse deliveries, a not insignificant item.

Sturdi-built Greenhouse Manufacturing Company
11304 S.W. Boones Ferry Road, Department BK
Portland, OR 97219
Phone: 1-800-722-4115
Product literature: catalog, $2

The American Classic Greenhouse

The American Classic series is the culmination of more than forty years of innovations in quality greenhouse design and construction by Texas Greenhouse Company. Sold through a nationwide dealer network, an American Classic greenhouse offers graceful curved eaves and an airy design that make it a splendid permanent addition to the home.

With their aesthetic appeal, the strength of these greenhouses, which comes from a galvanized steel substructure, is often overlooked. The combination of steel, extruded-aluminum glazing bars, and double-strength glass throughout produces one of the strongest, most attractive greenhouses offered at any price. Other American Classic features include full-length ridge vents with a special ball-and-socket hinge that will not be undone by high winds. Storm doors are self-storing and have screen panels that can be used for additional natural air flow.

Designed for use in virtually any climate, the wide range of the American Classic series, which includes both freestanding and lean-to models, puts ownership of a beautiful, efficient greenhouse within reach of most homeowners with a horticultural bent. These are not inexpensive greenhouses, but the AC-912, the smallest freestanding greenhouse in the series, has plenty of space for two benches to give the active gardener ample puttering room. The sturdy construction takes hanging plants in stride. Price: $3,554.

It's the mid-size AC-1500s, however, that are the most popular American Classic models. Priced from $6,280, for a unit measuring 15 × 15 feet, they are spacious enough for variations in bench arrangements to suit serious growing projects, have ample headroom,

The AC-1500s, the most popular American Classic models, may be purchased to fit on a kneewall or for full glass-to-ground construction. Standard lengths range from 15 to 37 feet. *(Texas Greenhouse Company)*

and plenty of walking-around area. Yet they're small enough to be easily maintained.

American Classic lean-tos are available in nine models and in many widths and lengths. They are made from the same materials and with the same attention to design as the manufacturer's freestanding greenhouses. Lean-to prices start at $2,545, for a unit measuring 6 × 12 feet.

All prices quoted are for mill-finish aluminum with clear, double-strength, 1/8-inch greenhouse-grade glass. Both painted aluminum and tempered glass (also tinted glass) are available as added-cost options. The colors most often ordered with the aluminum are architectural bronze and white, but any color can be matched at the customer's request.

Texas Greenhouse Company also sells the accessories needed to heat, cool, shade, and ventilate a greenhouse. Its dealers are also trained and experienced in greenhouse installation. Not all do-it-yourselfers have the exper-

American Classic lean-to greenhouses come in a wide range of roof pitches, lengths, and widths, provide a serious growing environment attached to the side of a home or other structure. *(Texas Greenhouse Company)*

tise for the concrete, plumbing, and electrical work that may be called for with installation of an American Classic greenhouse, and it is recommended that the purchaser of a greenhouse take advantage of the dealer's experience when possible.

Any glass structure needs a solid footing, and all American greenhouses are designed to sit on an 8-inch-high kneewall (24 inches in snow country), or can be ordered for full glass-to-ground construction at no additional cost. Plans are available, but the foundation or kneewall is the customer's responsibility.

Texas Greenhouse Company
2524 White Settlement Road
Fort Worth, TX 76107
Phone: 1-800-227-5447
Product literature: catalog, $3

Elite Greenhouses

The Elite greenhouse has been a top seller in England for more than twenty years. If price has been a deterrent to your dream, this classic glass import, from a country where one family in ten has a greenhouse, could be just what you've been looking for. At $999 for the standard 6 × 8-foot model, and $1,899 for the deluxe 8 × 12, here's quality at a hard-to-beat price. Among the Elite's features: sturdy, maintenance-free extruded aluminum framing; double-strength horticultural-grade glass; fingertip-controlled sliding glass door; adjustable roof vent; built-in gutters to carry water away from side walls while providing extra structural strength.

Assembly of a standard Elite greenhouse requires a screwdriver, an adjustable wrench, and a few hours' time. The larger unit will take a little longer. It is strongly recom-

Imported from England, aluminum-and-glass Elite greenhouses come in two sizes and at prices that the serious home gardener will find hard to beat. Shown here is the 8 × 12-foot deluxe model. *(Troy-Bilt)*

mended that the greenhouse be placed on a foundation of some type; pressure-treated lumber will do nicely. The predrilled aluminum framing, which includes diagonal members for extra strength, is assembled with screws. The precut glass panels slip easily into cushioned, "puttyless" weatherstripping and are anchored with stainless-steel clips.

There are a number of ways to ventilate the greenhouse. The sliding door can be left partially open, and every unit has a 2 × 2-foot roof vent (two in the larger unit) as standard equipment. At $50, an optional Set-and-Forget automatic vent opener would be a sound investment; as the inside temperature rises, so does the 2 × 2-foot roof vent, maintaining the interior temperature that you select. Other useful accessories include a louver window and automatic window opener. In warmer regions of the country, supplemental exhaust fans may also be needed.

It's the sheer volume of sales at home and abroad that have allowed the makers of the Elite greenhouse to hold down manufacturing costs and to sell at these low prices. To maintain that volume, Troy-Bilt, its distributor here, offers a no-down-payment purchase plan with monthly payments prorated for up to thirty-six months.

Troy-Bilt Manufacturing Company
102nd Street & 9th Avenue
Troy, NY 12180
Phone: 1-800-833-6990
Product literature: free brochure

A straight side-wall height of 60 inches allows ample room for hanging plants as well as placement of optional double staging tables and plant shelves in the deluxe model Elite greenhouse. *(Troy-Bilt)*

UTILITY SHEDS

If you live in a house with a big basement, a large attic, a two-car garage, and own only one car, finding storage space for the baggage of modern living may not be a problem. But in most households, storage space is at a premium, and it often comes down to parking the car in the drive because the garage is overflowing with bikes, lawn and garden tools and supplies, sports gear, barbecue equipment, and what most of us would simply describe as stuff.

Rather than move to a larger house, build an addition to the one in which you now live, or expand the garage, the most practical solution to acquiring more space, other than holding another garage sale or calling the junkman, is to install a utility shed or small storage building somewhere on the property. In kit form, these are relatively inexpensive, come in a wide range of shapes and sizes, and can easily be erected by the homeowner.

If utility sheds are common in your neighborhood, you might check with a few of your neighbors to learn of any problems they may have encountered either with the local building department or with their choice of shed. If a building permit will be required, pay a visit to the building department and tell them what you have in mind. There may be setback requirements and size limitations of which you weren't aware. The type and size shed could also determine whether a concrete slab or other foundation, as well as an anchoring system, will be needed. If you plan to install a power outlet or run water to the shed, electrical and plumbing permits may also be required.

Once you know the restrictions, if any, on the placement of the shed, and whether specific additional building criteria must be met, you can start making choices. How much do you want to spend? You can buy a 10 × 8-foot shed (a popular size) in galvanized steel for not much more than $200. You could also spend over $500 for a shed of the same size in wood. An easy decision? Not really. Steel rusts; wood doesn't. The wooden shed comes with a 15-year warranty. The metal shed is warranteed for only two years.

There are other differences. The steel shed has a gable roof with a peak height of under 6 feet and a door height of only 4 feet 10 inches. The wooden building has a barn-style roof line with a peak height of 8 feet, and a 6 × 6-foot door. The nondescript steel structure might look out of place anywhere but in a trailer park. The stylish, gambrel-roofed hut would win acceptance almost everywhere.

There are obvious as well as not-so-obvious differences in these storage buildings. A lot, though, can be read into the warranty. A metal building with a rust-resistant finish might carry a 5-year warranty. Metal with a vinyl coating could carry a 10-year warranty. Waferboard usually has a 5-year warranty. All of this is reflected in shed pricing. A gambrel-roofed, vinyl-clad shed can cost half as much again as a gable-roofed shed of the same size with a standard steel finish.

Metal sheds come precut, predrilled, and usually parts-numbered for ease in assembly. A Phillips-tip power screwdriver or a variable-speed drill with screwdriver and hex-head attachments can save a lot of time and effort here, since most of these kits are assembled with clips and screws. Wooden storage buildings usually come with preassembled wall and roof panels for rapid erection. Unlike with metal sheds, which are prefinished, the homeowner is responsible for such finishing touches as shingling the roof and staining or painting the exterior of the wooden structure. But then, it's these custom touches that put the wooden sheds in a superior class to most mass-produced metal sheds.

When it comes to finishing a wooden shed, exterior stains are easier to apply than paint. Latex-base stains offer easy cleanup, but we would recommend oil-base stains. These contain water repellents and preservatives, which protect against moisture-caused rot, splitting, and mildew. They can be applied with a brush, roller, or spray gun. Shingles could be chosen to match those used on the house itself.

You might not have a choice when it comes to a foundation, if one is needed for the shed. Local codes and practices will likely dictate the requirement here. This could be anything from a concrete slab or footings extending below the frost line to a gravel bed or a pressure-treated timber skid. With a wooden floor, a general rule of thumb is to provide 6 to 8 inches of separation between the finished grade and the bottom of the floor panels.

If the shed is to be used as a workshop or laundry room, a containment floor may be required to capture oil, grease, and other spills. If it is to be used to store a riding mower, a ramp will likely be needed. Do you plan to work in there in cold weather? If so, insulation may be desirable. A heating device, too. Most homeowners don't need electricity in a shed that serves as a storeroom. How often are you likely to go in there after dark? A flashlight is a lot less expensive than what an electrician will charge to run the wires and hook the shed to the household electrical supply system.

The air in a utility shed can get downright stuffy if the building doesn't get a chance to breathe now and then. If the kit doesn't come with one, give serious consideration to installing a vent or window in the rear wall or a wind turbine on the roof of the shed.

Handy Hut Wooden Storage Buildings

We've already expressed our preference for wooden rather than metal storage buildings, and the popularity of the Handy Hut line of ready-to-assemble kits shows we're in good company. If you're looking for a utility building that will be at home in any setting, Handy Hut offers attractive wooden storage buildings in a wide choice of styles and sizes.

With these kits, there's nothing to saw. All you need is a hammer, screwdriver, and tape measure. Every kit comes complete with preassembled doors, walls, and roof panels, plus pine trim, hardware, and step-by-step, illustrated instructions. A videotape covering the

Textured plywood siding, reinforced crossbuck doors, and extra-wide trim enhance the "barn-style" look of the Heritage, which is the most deluxe of all Handy Home wooden storage buildings. *(Handy Home Products)*

The 8 × 8-foot gambrel-roofed Premiere is the most popular and, at a kit price of $260, carries the lowest cost per cubic foot of all Handy Huts. An assembly video is included with this kit. *(Handy Home Products)*

assembly is also included in several of the kits to make installation a breeze. The videos show everything from clearing the ground and building a floor to shingling the roof and adding vents. You need no special carpentry skills to assemble any of these kits.

The most popular and lowest in cost per cubic foot of the Handy Huts is the 8 × 8-foot, gambrel-roofed Premiere model, with many thousands providing extra storage space in yards across the country. Siding and roof panels of the economy-priced Premiere are durable waferboard. Peak height is 8 feet, and the door opening is 4 feet wide and nearly 6 feet high. Estimated assembly time is 3 to 5 hours. The hut comes with an assembly video and a 5-year warranty. Kit price is $260. The structure must be erected on a wooden floor system or concrete slab per instructions. A floor kit of treated lumber and waferboard is offered as an option for $100. Finishing touches require the purchase by the homeowner of four bundles of roof shingles, a gallon of exterior paint or stain for the sides, and a quart of another hue for the trim.

The top-of-the-line Handy Hut is the Heritage. Quality rough-sawn fir-plywood siding, reinforced crossbuck doors, and extra-wide trim enhance the look of this "barn-style" hut with traditional soffit overhangs. The Heritage is available in four sizes, 10 × 8, 10 × 12, 10 × 16, and 10 × 20 feet, all with an 8-foot peak height. The kits come with an assembly video and a 15-year limited warranty. Estimated assembly time for the 10 × 8 structure is 4 to 7 hours. Kit prices for the Heritages range from $530 to $1,300. Floor kits are available from $125.

The Classic, constructed with waferboard siding and roof panels, is a simpler version of the 10 × 8-foot Heritage. It comes with an assembly video and a 5-year limited warranty, for a kit price of $400. Rounding out the hut line, the 8 × 8-foot gambrel-roofed Seville features rough-sawn channel-groove plywood siding, reinforced crossbuck doors, and soffit

Step-by-step detailed illustrated instructions with each kit.

overhangs. It's more or less a deluxe version of the 8 × 8-foot Premiere. It, too, comes with a video, and has a 15-year limited warranty and a kit price of $360.

The line also includes two "lean-to" designs that can be placed against a garage or house wall. Extra-wide double doors offer easy access to the contents in both the 6 × 4-foot Aspen and the 8 × 4-foot Franklin, which are kit-priced at $210 and $270, respectively. There's also the free-standing, slope-roof Pioneer, which measures 6 × 4 feet and is a convenient size for the smaller yard. With the economy of waferboard siding and roof panels, the Pioneer has a kit price of $200, floor kit included. All three units have a 7-foot peak height.

Also worth a mention here is the Handy One Recycling Center and Refuse Organizer. For indoor or outdoor use, the Handy One is made with textured plywood siding, measures 6 feet wide, 4 feet high, and 2-plus feet deep. Lift-up tops and space-saving sliding front doors allow easy access to, and removal of, any of the four plastic bags, which are held in place with metal hoops. There is a 6-foot-long, 10-inch-deep shelf inside the organizer and one on top. Kit price, floor included, is $180. Assembly time is 1 to 2 hours.

Any of these kits, which are produced by Handy Home Products, can be ordered through Sears, Scotty's, and other home centers.

Handy Home Products
P.O. Box 548
Walled Lake, MI 48390-3222
Phone: 1-800-221-1849
Product literature: free Wood Buildings brochure

Green Mountain SunShed

Here's a sturdy outdoor storage shed you can use in so many ways! It's a sunny potting shed, a mini-greenhouse, a clean, dry storage space for tools and outdoor equipment, a garage for the kids' bikes, sleds, skis, and other gear. It could also make a fine little workshop for pursuing a craft or hobby or writing the great American novel.

Much more than just a garden shed, the Green Mountain SunShed was designed by an architect and exceeds building codes for new-home construction in the snowbelt of Vermont. Framing is kiln-dried fir and stress-rated spruce 2×4s. Flooring, siding, and roof

The saltbox styling and sound construction of the multipurpose SunShed make it an attractive addition to any yard or garden. *(Gardener's Supply Company)*

sheathing is ½-inch or better fir plywood. With its handsome New England saltbox styling, it can make an attractive addition to any yard or garden.

The SunShed has a peak roof height of 8 feet and comes in two sizes, 8 × 8 feet and 8 × 10 feet. Kit prices are $1,845 and $2,195, respectively. Roof shingles and paint or stain are the owner's responsibility. With preassembled and predrilled floor, wall, and roof panels, assembly (covered in a 16-page SunShed Owner's Guide) is as easy as one-two-three.

No permanent foundation is needed (unless required under local building codes). Any dry, level location will work. The reinforced flooring can be placed atop landscape timbers or a similar elevated base. Panels are joined together using the galvanized lag bolts and other hardware provided. After the roof has been added, the greenhouse window is slid into place and fastened. With preassembly and a leakproof glazing system, it takes only minutes to install the window. Pine trim completes the SunShed assembly. It's then up to the owner to add roof shingles and to stain or paint the structure to match or complement the home.

Other finishing touches might include the addition of a window box, landscaping around the base, and a rose bush trained to climb up a wall. The interior of the SunShed can be custom-designed with the addition of shelves, tool racks, a potting bench, or whatever is required to suit the family's needs.

A hinged gable window that may be locked open in several positions and a Dutch door allow flow-through ventilation. If power is available, a fan, regulated by a wall-mounted thermostat, could be installed in either gable end to facilitate automatic venting.

For those who might be short on imagination, the SunShed product literature includes a rundown on "101 Ways to Use a SunShed Year-Round." For qualified buyers, the SunShed can be purchased with no money down and low monthly payments.

Gardener's Supply Company
128 Intervale Road
Burlington, VT 05401
Phone: 1-800-533-4551
Product literature: free Gardener's Supply Company catalog

The Throne

The Throne is a modular version of the classic outhouse or Chic Sale. Introduced by The WoodsPeople of Maine to go along with its line of panelized shelter systems, most of which are erected for use as economical second homes, it's finding its greatest popularity with buyers who are adapting it for tongue-in-cheek use as a utility building or garden shed, or as a poolside changing room.

The 4 × 4-foot Throne comes ready to assemble for $495, FOB, Bangor, Maine. It's an easy bolt-together design built with standard plywood panels. To avoid freight charges, most out-of-staters choose to purchase the wood locally and build The Throne from plans, which can be purchased separately for $15.

The WoodsPeople have been active in designing, improving, and fabricating wood shelter systems for more than twenty-five years. They're good people to know. Another of their recent innovations is the UniShed, which can be built in 4-foot increments around a newly designed frame system. The WoodsPeople provide the "exo-skeleton" and hardware. The buyer provides the plywood, roofing, and labor.

The Throne might raise a few eyebrows but it can easily be adapted for use as a novel utility building or garden shed. *(The WoodsPeople)*

The WoodsPeople
19 Freedom Parkway
Bangor, ME 04401
Phone: 207/848-7200
Product literature: information package, $3

PLAYHOUSES

Most of the kids' playhouses you see in the stores today come in either not-so-indestructible plastic or corrugated cardboard. But here's the real thing—a substantial (shipping weight is 290 pounds), attractive cedar chalet that will provide years of fun for your children and theirs. The wood weathers naturally if left unfinished but can be stained or decoratively painted. It's another easy-to-assemble, all-wood outdoor structure from Handy Home Products (see pages 94 and 95).

The Cedar Chalet is 6 feet long, 4 feet wide, and 6 feet high. Its features include side and front windows, a flower box for the front window, a stylish panel door, and a porch with railings and balusters. The roof covering is bevel siding and the floor is cedar decking. The kit comes with major sections preassembled and requires only 2 to 3 hours to erect. Kit price is $380.

The playhouse can also be ordered from Grand Era Reproductions (P.O. Box 1026, Lapeer, MI 48446; phone: 313/604-1756). Grand Era adds decorative brackets and gingerbread fretwork to give the little house a charming Victorian accent. Their price for "Katie's Playhouse" is $469. Grand Era also offers a small cast weathervane—rooster or running horse—as an optional roof decoration for $22. A scaled-down picket fence with an operating front gate, to provide a finishing touch for this special child's playhouse, is also offered.

The captivating Cedar Chalet has the built-in durability to provide years of fun for your little ones and theirs as well. *(Handy Home Products)*

DOGHOUSES

Made with durable plywood construction, Kanine Kabins have only eight parts to assemble, come in three sizes, and are priced from $38. *(Handy Home Products)*

A doghouse, even if you don't own a dog, can be a good investment in these days of rising crime. Criminals look for easy victims and could have second thoughts about invading your property if they know, or even think, that there's a hound around. And the bigger the doghouse the better.

If you do own a dog, or are thinking about acquiring one and would like to keep it outside at least part of the time, your pet deserves a home it can call its own. It should be large enough for the dog when full grown to lie in without touching the walls.

The trend in doghouses seems to be toward easy-to-clean, one-piece molded plastic structures. But we remain unconvinced that they would be the doghouse of choice with every canine on the block. For an alternative choice, our friends at Handy Home Products (see pages 94 and 95) offer the more traditional, all-wood Kanine Kabin.

The gable-roof Kabins come in three sizes: medium, large, and extra large. They are made with quality rough-sawn plywood and feature factory-applied shingle-type roofing, a circle-top door opening, and an elevated floor to keep Rover and bedding off the ground. There are only eight parts to assemble, with just a hammer. One addition we would make to the Kabins would be a weather flap, to ensure near year-round comfort for our deserving friend.

The large-size Kanine Kabin measures 24 × 32 × 30 inches, with 30 inches being the height. It's for dogs weighing between 40 and 70 pounds. The listed price for the kit is $48, but the price you pay may vary depending on where it's purchased. The medium-size Kabin carries a $38 tag. The extra-large Kabin, which measures 32 × 48 × 32 inches, has a list price of $65.

Man's best friend is deserving of the best, and some owners go all out and provide custom-made quarters for their canines that are mini-replicas of the pet owner's own home. There's nothing in kits to equal that. But if you are looking for something unusual, the design and engineering staff at Almost Heaven (see pages 132–34) say they can create anything you fancy, and offer the "barrel house," shown here, as proof.

Looking for something unusual? Almost Heaven, the company that makes the barrel sauna, offers this conceptual "barrel house." *(Almost Heaven, Ltd.)*

BIRDHOUSES AND FEEDERS

In the late 1800s, it was not uncommon for architects to include birdhouses as tiny outbuildings replicating the main house. Made with tin, wood, and even cast iron, those vintage, one-of-a-kind birdhouses have become highly prized as collectors' items. There *are* kits today that are as much birdbrained whimsy as those earlier conceits. But if you wish to attract certain avian species, whether for their song, color, or good deeds, you really need a house that meets Audubon specifications for that bird.

Take bluebirds, which are making a comeback after DDT and loss of nesting habitats virtually wiped them out. They are in need of houses. But to protect this lovely songbird and its nest from predators and its primary enemy, the starling, the houses should have a deep cavity and an entrance opening no greater than 1½ inches in diameter. Bluebirds also appreciate a rough-sawn interior. Wrens, on the other hand, are one of the easiest of all birds to attract to man-made houses of any design. But if the opening is greater than 1 inch in diameter, nest-stealing house sparrows and other less-welcome species are likely to dispossess the tiny wrens.

Communal birds like the insect-devouring purple martin need an altogether different type of housing, and a whole industry has developed to provide multicompartmented martin houses to attract what has become the gardener's favorite and one of America's best-loved birds. The typical martin "condo" is aluminum, with a heat-reflecting baked-on-enamel finish, and comes with a mounting post on which it can be raised or lowered like a flag for easy maintenance.

Serious bird watchers can tell you all about the nesting and feeding habits of a wide variety of bird species. They can tell you which birds are attracted to man-made houses, which will reward you by devouring insects, and which are likely to pass you by because you've filled a feeder with their least favorite seed. In other words, you can put up a birdhouse and set out a feeder and take pleasure in whatever comes along, or you can seek to attract specific bird species.

Apart from martin houses, most birdhouses built from kits are inexpensive enough that you might put up two or three different types, though well-separated, and keep note of what interesting and colorful birds each house attracts.

Birds also require clean water, for drinking and bathing, and a source of food, or they're not likely to hang around for long. Again, certain feeders, as well as seed, attract certain species. The most common feeders in use today are the hanging-tube type made with plastic. But it's the feed that ultimately attracts the birds, so the more traditional wooden shelf-style feeders offered in kit form still have their place.

Thistle seed is useful for attracting goldfinches, and cracked corn and peanut hearts are enjoyed by cardinals, but it's black-oil sunflower seed and shelled sunflower meats that will attract the greatest variety of watchable birds. We don't recommend wild birdseed, since it contains millet, which attracts house sparrows and starlings, both of which can spell trouble for the more rewarding cavity-dwelling birds.

Don't be disappointed if birds don't appear at once. It could take a number of weeks before they find your feeder, and even longer before they set up house in a new home.

Woodkrafter Kits

If you are looking to attract bluebirds, chickadees, or wrens to your yard or garden, a Woodkrafter birdhouse kit could be just the attraction you need. Made from precut, predrilled cedar parts, with no finishing required, Woodkrafter kits meet Audubon specifications. They're educational projects, too, suitable for youngsters ages eight and up, or for an adult and child working together.

Made from precut, predrilled cedar parts, Woodkrafter birdhouse and feeder kits meet Audubon specifications and make great educational projects, suitable for youngsters ages eight and up.

To assemble the birdhouse kits, all you need is a hammer. Nails are provided. There's also a bird-feeder kit, for which no tools are needed; it features pegged construction, requiring only household glue. The cedar for the birdhouses is half an inch thick and naturally weather-resistant. With some species, though, the more weathered the wood, the more they seem to like it.

The deep cavity and the 1½-inch opening of the bluebird house make it particularly inviting to this species. To assemble the birdhouse, all you need is a hammer. Nails are provided.

Bluebirds, in particular, like rough wood. Sanding is not recommended. With the deep cavity required to keep nestlings beyond the reach of cats, raccoons, and other predators, a rough-sawn interior provides the birds and their young the traction needed to climb to the opening.

A lot of useful information about the species is included with each kit. For example, you learn that the female bluebird lays four to five eggs, and that young birds can leave the nest two to three weeks after hatching. So that the fledglings have a safe place to which to fly, the birdhouse should be located to face a tree or large bush no more than 30 to 50 feet away.

It also helps if the house is situated where it can be easily reached by the homeowner after each nesting, or after the first brood has left. Bluebirds are early nesters and may raise one or two additional broods if the nest is cleaned out between broods. To this end, the kits are designed with a removable bottom for easy cleaning.

The house for the perky little wintertime chickadee is designed much like the deep-cavity bluebird house, but with a considerably smaller entrance hole. The predrilled hole for the chickadee house measures 1 1/8 inches in diameter, versus the 1 1/2 inches specified for the bluebird house. The entrance to the gable-roofed wren house is even smaller. It's only an inch in diameter, which serves to deter the pesky but somewhat thicker-waisted house sparrow.

No tools are needed to assemble the bird feeder. It features pegged construction, requiring only household glue. We've filled it with sunflower seed, the favorite treat of most songbirds.

Woodkrafter's feeder kit makes a gable-roofed, hopper-style feeder, with feeding troughs on both sides of the clear-plastic-faced hopper. If necessary, birds can be lured to the feeder by placing feed in a shiny pan placed beneath the feeder until the feeder itself is discovered. This is likely to attract squirrels, but if the feeder is hung 4 to 5 feet above the ground with the cord provided, and away from the tree trunk or a thick branch, squirrels will be discouraged from trying to get to it. Keep the feeder clean; old moldy seed can endanger birdlife.

Woodkrafter kits are sold through hobby shops, catalog houses, and other retailers. We purchased our bluebird house ($9.75) and bird feeder ($9) kits from The Brown Company (Yawgoo Pond Road, West Kingston, RI 02892; phone: 1-800-556-7670). The Brown Company, a supporting member of the National Audubon Society, offers a wide variety of items, including books, records, and videos, directed at encouraging the protection of wild birds. Their catalog is free.

Woodkrafter Kits
42A North Elm Street, P.O. Box 808
Yarmouth, ME 04096
Phone: 207/846-3746

Greenleaf Products

Not all nesting birds will take to Greenleaf's fanciful birdhouses. But those who do will have something to yodel about. For they're a far cry from nesting in something as primitive as a cavity in a tree. The motif of Greenleaf's birdhouses is Swiss/Bavarian. To be right at home, the feathered residents really should be tricked out in tiny dirndls and lederhosen.

Greenleaf's wooden birdhouse and feeder kits are die-cut from exterior-grade plywood and feature interlocking construction—requiring no nails, screws, or glue. They're easy to build and fun to decorate. A painting pattern is included in every kit, but there's no reason that you can't apply a design or colors of your own choosing. Tole and decorative painters love the little houses and often become so taken with their handiwork that the finished abodes end up as a home decor item.

The die-cutting leaves the edges of the precut pieces a bit ragged, but sandpaper is included in the kits to smooth whatever needs

The motif of Greenleaf's birdhouse kits is Swiss/Bavarian. A painting pattern is included in every kit, but there's no reason you can't apply a design or colors of your own choosing.

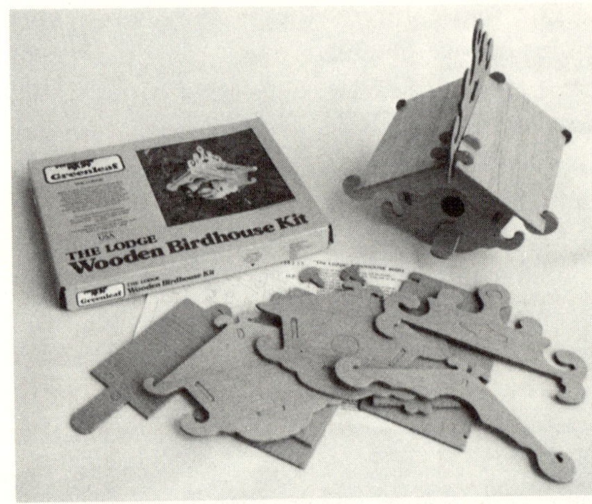

Greenleaf's kits are die-cut from exterior-grade plywood and feature interlocking construction—requiring no nails, screws, or glue. Sanding is necessary, however, to smooth most edges.

to be smoothed. If the structures are to be exposed to the elements, though, we strongly recommend that all edges and exterior surfaces be sealed with a minimum of two coats of marine spar varnish or other exterior primer before the final weather barrier is applied.

If you plan to follow the painting pattern provided with the kit, you'll find it's easier to apply the sealer and transfer the design while all the pieces can still be laid out flat. For a less complicated painting scheme, it might be preferable to apply the finishing touches after the structure has been assembled. One additional step that we recommend is to seal the ridgeline with a wood filler. If this is not done, nests and seed could get wet when it rains and introduce unnecessary health risks.

Greenleaf, best known for its die-cut dollhouses (see pages 237–42), offers two birdhouse kits and one bird feeder. Its cozy little Lodge is designed with a 1-inch opening to attract wrens, finches, and other small birds. Its larger Swiss Inn, with a 1½-inch opening,

The Lodge is designed for wrens and other small birds, the Swiss Inn for larger birds like the cardinal, while the Black Forest's feeding troughs should attract a variety of watchable species.

could attract cardinals. Its Black Forest bird feeder includes sturdy plastic windows through which the level of seed can easily be seen. The kits, which can be purchased at home centers, hobby shops, and elsewhere, retail for between \$10 and \$12. Paints are not included.

Greenleaf Products, Inc.
58 North Main Street, P.O. Box 388
Honeoye Falls, NY 14472
Phone: 1-800-847-2545

Nature House

The purple martin, largest member of the swallow family, eats only flying insects. It can easily tuck away two thousand or more mosquitoes a day if available. Which is a good enough reason that the martin has become America's most wanted bird. But in addition, it's also a beautiful, graceful bird, and an unusually interesting one. Unlike most birds, martins appear to be curious about the human species, and are aptly termed "semidomesticated."

The only way to attract martins, however, is to put up a house for them. North American Indians learned this centuries ago and hung up dried, hollowed-out gourds to lure the insect-eaters to their pest-ridden campsites. In 1962, at a time when the martin was being considered for the endangered species list, Nature House started a crusade for them and began development of innovative housing that would benefit both the martins and their landlords. The resulting Trio aluminum houses and systems have been a major factor in one of this century's more successful conservation stories.

Purple martins are most prevalent in the eastern half of the United States. They're also

The Trio-Castle provides a home for twenty-four pairs of martins and their broods. Wide, wraparound porches and a wheel-spoke rooftop perch accommodate those friendly martin gatherings. *(Nature House, Inc.)*

found along the West Coast and sometimes in western mountain localities. The birds winter in Brazil, and each colony sends out a scout to find a home before returning to North American locales early in the year. When they find what they like, purple martins are inclined to return year after year. Typical colonies range from a few pairs to twenty or thirty pairs. Some colonies have more than one hundred pairs.

The Trio line of all-aluminum martin houses includes models with six, eight, twelve, and twenty-four compartments. The 12-compartment Trio-Grandpa (TG-12), the world's most popular martin-house system, raises and lowers like a flag for easy maintenance. The biggest, most impressive house in the line is the Trio-Castle, a four-level hexagonal house with twenty-four compartments and a wheel-spoke rooftop perch for those friendly martin gatherings. The kit includes a sectional steel post and a winch-and-cable assembly that allows the Castle to be raised or lowered with just two fingers. The package also includes

The 12-compartment Trio-Grandpa is the world's most popular martin house. It raises and lowers like a flag, with a nylon lanyard, and is the easiest of all Trio houses to maintain. *(Nature House, Inc.)*

removable Dri-nest subfloors and winter door stops (to prevent sparrows from gaining access when the martins are gone).

The Trio systems range in price from around $125 for the single-level, 6-compartment house, with a telescoping mounting post (adjusts from 5 to 14 feet) and a ground socket to hold the post, to $330 for the 24-compartment Castle, with post, winch, and ground socket. All units feature porches, with guard rails to keep young martins from falling to the ground before they're ready to test their wings. The green-and-white baked-enamel finish on the houses is heat reflective and the bright aluminum interiors discourage starlings, which prefer dark cavities. Flip-up fronts and lift-out subfloors make evicting squatters and annual fall clean-out easier.

In addition to mosquitoes, martins consume vast numbers of gnats, flies, and other insects. They forage on the wing over a wide area, and their acrobatic antics are a joy to watch. The houses should be positioned in an open area at least 15 to 25 feet away from tall trees or buildings that could interfere with the bird's swooping approach to the nesting site.

Trio purple martin products are sold through hardware and general merchandise chains; seed, lawn-and-garden supply, and general merchandise catalogs; and independent dealers. If you can't find a retailer in your area who handles Trio products, write or call Nature House for the name of the dealer nearest you.

For $12 a year, you can also subscribe to the monthly *Nature Society News,* which for over twenty-five years has been "The Voice of the Purple Martin." It gives coverage to bluebirds, hummingbirds, and other ornithological favorites as well.

Nature House, Inc.
Purple Martin Junction
Griggsville, IL 62340
Phone: 217/833-2393

PLAYSETS

Children learn through play, and there's nothing like a developmentally appropriate backyard playset for stimulating their imagination while at the same time providing activities for physical and social development. You really need to provide a variety of activities. Unless interchangeable play components are available to suit growing skills, a swing set alone is not going to do all that much for the child.

Don't buy on the basis of price. With some of the less-expensive playsets, you're shortchanging not only your child but yourself. The cheaper structures made from steel piping can deteriorate quickly and become hazardous with heavy use. Look at a playset as an investment in your children's future. It should not only be sturdy but hold the promise of being both eye- and child-pleasing until your last-born has outgrown its attractions.

They may cost more, but a wooden play structure built with pressure-treated 4×4s and 4×6s can last twenty years or more. If you are concerned that your children will outgrow it long before its time, there are systems that lend themselves to easy conversion to a "universal gym" or family fitness center. If yours is a young family and you're on a tight budget, you could start with a basic system that includes a couple of swings and a plastic slide, and add to it later. Most of the better playsets are easily expandable. The cost for a good starter set will run from $350 to $550, depending on whether the structural lumber comes with the kit or is purchased locally. You could also go all out and spend a couple of thousand dollars on a backyard extravaganza with everything an army of preschool and older youngsters could want.

Costs and accessories aside, swings and playsets should be located on level ground not less than 10 feet from any structure or obstruction, such as a wall, laundry lines, or overhanging tree branches. Be sure to allow for future expansion if a slide and climbing tower or the like are not included as part of the initial system. To ensure safe play, the surface under and around all of the equipment should be soft and resilient.

Some playsets are freestanding, others come with supports that must be anchored in concrete. Where concrete footings or anchors are called for, keep them below grade and covered over with several inches of impact-absorbing turf. Supports and beams should be joined with nuts and bolts, either recessed or with ends capped in such a way that they pose no hazards to the active young. Corners and edges of all wooden structural members should be rounded and sanded smooth.

For safety's sake, don't buy a swing seat beyond your child's years. There are structural-foam seats with leg holes and a safety belt for ages six months to two years; wrap-around supports with a front belt and snap for children ages two to four; and, for older children, the reinforced-vinyl sling or belt seat. Good sling seats cost about $20, infant bucket seats, $25, and in-between seats, about $35. Stay away from wooden and metal seats.

There's also been a turn away from wooden and metal slides. The slides of choice today are plastic safety slides, with high sides and an exit gradient that doesn't send the child tumbling to the ground. The molded polyethylene slides, usually a fade-resistant bright yellow, stay reasonably comfortable even on days that can turn a metal slide blistering hot. Plastic safety slides are priced between $100 and $120 for an 8-foot slide. They also come in longer lengths.

The playset should match the ages and skills of those who will be using it. If you have youngsters several years apart in age, you may need to separate the centers of activity, with a baby swing and slide or sandbox at one end and more challenging climbing and swinging devices at the other end. To provide the essentials of active play for growing children, you may, either now or later, want to consider adding such options as a climbing rope, a slide pole, monkey bars, or some of the other accessories pictured in the pages that follow.

For safety's sake, with any playset, no matter how simple, on-site adult supervision must be provided for children under four years of age. You also owe it to the family to check the playset frequently for any loosened hardware, splintering, or rope wear. Tighten, sand, and replace—and you can keep accidents from happening.

BigToys

If you have children under school age, you may already be acquainted with BigToys creative playground equipment. It has been used to provide more than fifty thousand successful outdoor play structures in day-care centers, preschools, and neighborhood parks. What you may not know is that with Backyard BigToys, designed for residential settings, you can provide a similarly rich and diverse play environment right at home.

BigToys play structures get their strength, stability, and good looks from lodgepole pine logs milled to a smooth, 6-inch diameter. The

With plastic slide, swings, rope climb, slide pole, monkey bars, a BigWheel, and stout panels with plenty of handholds, the Aspen is designed for an age range of three to nine, and costs $999 in kit form. *(BigToys)*

Kids love to make believe, and this Backyard BigToys Signature series structure designed by Sami Bandek allows them to travel to all sorts of fantastic, faraway places. Kit price is $1,899. *(BigToys)*

logs are also center-bored to prevent splitting from drying and are pressure-treated inside and out with environmentally safe Wolman preservative. They provide a durable and attractive alternative to the piping used in the A-frame swing sets sold by the mass-merchandisers.

With its variety of modular play centers and a wide range of options, BigToys makes it easy to choose or create a design that matches the needs and capabilities of your children. The Magnolia system, for example, with a baby swing, a belt swing, and a 6-foot plastic slide, is designed for ages ranging from two to five years. Its highest deck is 24 inches. Kit price is $549, and assembly time is approximately 6 hours.

The massive Sequoia, on the other hand, is designed for ages from four to twelve. It accommodates younger children while at the same time continuing to entertain and challenge older brothers and sisters. It features a tall, two-level tower, and its standard play elements include an 11-foot-4-inch plastic slide, a baby swing, two belt swings, a rope climb, a slide pole, and monkey bars. Its highest deck is 66 inches. Kit price is $2,099 and assembly takes approximately 18 hours.

Every detail of these structures has been designed from the perspective of the children using the equipment. Decks spiral up at 12-inch increments. One-inch-diameter bars are easy for small hands to grasp. Panels feature ideally spaced safety-grip handholds. A free-spinning BigWheel is a catalyst for open-ended fantasy play. The slide pole and knotted climbing rope provide just the right amount of physical challenge for preschoolers.

Options offered by BigToys include everything from gym rings and a trapeze bar to the Polynet (a more challenging access to the play structure, it improves eye-hand coordination and makes a great hammock for lounging) and a spiral slide. Any of these items can be purchased separately, either to add to an existing BigToys system or for use with a playset built from stock lumber by a do-it-yourselfer.

When your children outgrow their BigToys play center, many of the components can be used in an outdoor universal gym or family fitness center for ages eight and up, extending the life of your original purchase. Adults can tone up on the wide range of strengthening activities while helping future Olympians reach their fitness potential. BigToys also sells a complete Log Gym system, featuring eight exercise stations. It has a kit price of $1,199.

BigToys
7717 New Market Street
Olympia, WA 98501
Phone: 1-800-426-9788
Product literature: free catalog

Spiral slides are a recent innovation but have already become a favorite play activity across the country. Your children will never forget the day this is installed in your backyard. *(BigToys)*

WoodBuilt

Does it make sense to truck lumber across the country to construct a backyard play center when equally fine products are available locally? WoodBuilt of Wisconsin provides a money-saving alternative with do-it-yourself kits that include plans, mounting hardware, fasteners, swings, ropes, and ladders for a variety of play systems. Give your local building supply the list of lumber needs, included with each kit, and you are in business. You not only save on wood finishing and shipping costs but are able to choose from pressure-treated lumber, redwood, and cedar.

These playsets use 4 × 4s and 6 × 6s for extra strength and durability, and are designed and described to simplify the job. They are an easy weekend project, and you even get cutting guides for the lumber and plenty of illustrations. There's no digging or concrete work involved; each unit is freestanding. You will appreciate the heavy-gauge brackets and avoidance of miter cuts. All components are designed to meet or exceed current guidelines for backyard activity centers.

WoodBuilt's easiest-to-build playset is the Scout, a wooden A-frame swing set that allows room for expansion. The Scout kit costs $99.50 and includes four EZ-Frame

WoodBuilt's easy-to-build Scout play center makes an ideal starter set for the young family. Estimated cost, including locally purchased lumber and the optional EZ Ryder, is $225. *(WoodBuilt)*

Shown here with WoodBuilt's swinging bridge and merry-go-round options, the all-in-one Pioneer play center (kit price, $209) offers plenty of swinging, sliding, and child-pleasing climbing. *(WoodBuilt)*

braces, four EZ-Frame brackets, three pairs of swing hangers, two sling-type swing seats and a trapeze bar with "ouchless chain," and the hardware mounting kit (bolts, nuts, and washers). The estimated cost of the lumber required is $34. Swings are well-spaced, and there's a fourth play area left open for the installation of an additional accessory of your choice. The Scout makes an ideal starter set for the young family. Later, as the youngsters develop, climbing towers and other features can be added. Suggestions for future expansion are included in the booklet of instructions.

The Eagle's Nest Plus, which features a "lofty" clubhouse, is WoodBuilt's most popular activity center. It can be constructed as a single-level clubhouse or a four-level climber. A 28-page fully illustrated guide helps you put either of these sets in your backyard. In addition to the necessary hardware, the kit includes a climbing rope, a rope-and-dowel climbing ladder, a blue vinyl tarp for the clubhouse roof, and an 8-foot Cool Wave Slide. The Eagle's Nest Plus kit costs $219. Estimated cost of the lumber required is $139.

Lots of climbing and lots of swinging best describe the Fort Eagle, WoodBuilt's new top-of-the-line activity center. With swings, rings, monkey bars, slide, ladder, climbing rope, and climbing cargo net, this 24 × 20-foot play center will exercise every young muscle.

The Sky Fort, a backyard extravaganza, has everything an army of youngsters could want, including a clubhouse with roof, slide, climbing rope, cargo net, slide pole, and a merry-go-round. *(WoodBuilt)*

Hardwood dowels are supplied for the ladder and monkey bars. The kit costs $359, and the estimated cost for the lumber is $213.

With WoodBuilt kits, it's always a good idea to leave room for expansion. Any of these play centers can easily be enlarged or modified to meet changing family needs.

WoodBuilt
One Parker Place, P.O. Box 92
Janesville, WI 53545
Phone: 1-800-475-5051
Product literature: free brochure

HOT TUBS AND SPAS

The therapeutic benefits, as well as the mind-freeing pleasure, to be derived from soaking in hot water, whether still or bubbling, were known to many ancient cultures. In earlier times, hot tubbing and spa bathing necessitated a trip to a communal bathhouse, or more distant travel to a "hot spring." Today, for less than the cost of the average family vacation, you can install a hot tub or spa in your own backyard and share the joy with family and friends for many years to come.

Hot tubs come in diameters of from 4 to 8 feet, and in depths of 2, 3, or 4 feet. The most popular tub size is 6 × 3 feet. Tubs of this size can hold four to five adults comfortably and can be purchased for under $1,250 in Western red cedar, or $1,600 in redwood. Add the cost of a heater or a spa pack, plus a tub cover and other accessories, and you're still looking at an outlay of only $2,500 to $3,500.

You won't save much by going to a smaller tub, since the cost of the mechanicals is pretty much a constant. But if the budget's tight, you could start with just the tub, or a tub and heater, and add the spa components at a later date. You might even find there are benefits enough in just a nice long soak, and have no call for bubbling water or multilevel hydro-massage jets.

A hot tub needs a firm, level foundation. This could be provided by a patio or a well-supported deck. Otherwise, tub location is not critical unless there's a need for privacy, or to make it less accessible to young children. In either case, you might want to locate the tub behind a fence or other barrier.

If there *are* young children about, the tub should never be left uncovered in the absence of proper supervision. We're talking about a solid, securable cover, and not just a floating heat cap. A form-fitting, insulated cover with a respectable R value will more than pay for itself in reduced heating costs if the tub is to be kept in use during the colder months. It will also serve to keep falling leaves and other airborne debris out of the tub.

You'll need a heater, of course. For without hot water, it wouldn't be a hot tub. Gas heaters work faster than electric heaters in raising the tub temperature, but they're also more expensive. For a 6 × 3-foot tub, a quality electric heater will cost between $500 and $650; a comparable gas-heating system can easily cost $1,000 or more. As an alternative, you could install a wood-fired stove (about which, more later) for $625. They can heat up tub water fastest of all.

Hot tubbers who prefer to keep it simple change the water every few uses and don't bother with the purifying chemicals or filters that others rely on to maintain proper water chemistry. For longer periods, however, there *is* a need to chlorinate the water. Filters may also be needed to keep the water from turning cloudy. Hassle-free filtration requires a pump. And if you're buying a pump, you'll probably want a two-speed pump. It runs slow for circulating and filtering, and to keep the water from freezing in cold weather. Fast speed is for jet-massage action.

A jet/filter system for a hot tub will add another $600 or more to the cost of the system, which is now more spa than hot tub. The next step would be to add hydro-massage jet heads and suction fittings. But rather than build the system piece by piece, you can get it all in a spa pack that includes an electric or gas heater, a two-speed pump, and a cartridge filter—all preplumbed and mounted on a single pallet. Almost Heaven offers quality electric and gas spa packs priced from $1,660.

Benches are usually included as part of the hot-tub package. They go inside the tub. For dignified access to the tub, you'll need a set of stairs or box steps, which can be purchased as an accessory. For a more deluxe treatment and to expand the capabilities of the tub, you might want to consider a precut tub surround. You could also construct a partial surround or deck and steps yourself from locally purchased materials.

A 24-hour timer, available as an accessory from most tub manufacturers, can be programmed to start up the tub system and have the water at the preferred temperature for regular morning or evening dips. As the National Pool & Spa Institute points out, "hot" should not be taken too literally. The maximum safe tubbing or spa temperature is 104°F. (40°C.). Fifteen minutes per soak is the recommended maximum time at that temperature. If you're inclined to take longer soaks, cool it.

You might also want to invest in a freeze protector. But the best protection here, if a wooden tub is going to be left unattended for more than a week, is to drain all but 3 to 4 inches from the tub and let the remaining water freeze. The ice will help to keep the tub "seasoned" and won't do any harm. This advice is from folks who should know. Their hot-tub business got its start in Alaska.

Almost Heaven

There's no place like Almost Heaven when it comes to one-stop shopping for hot-water health and leisure products. Its hot tubs alone come in eight standard sizes and are made in

With Almost Heaven's hot-tub systems, which come in eight standard sizes and are made in four different woods, you can have the beauty and durability of real wood and the convenience of a spa. *(Almost Heaven, Ltd.)*

four different woods, beginning with redwood, which is where most other manufacturers leave off. Almost Heaven also offers spas, saunas, steamrooms, and whirlpool baths, as well as a full line of tub and spa accessories.

The West Virginia manufacturer's hot tubs, in a choice of California redwood, Honduras mahogany, Burmese teak, or American white oak, are packaged three ways:

(1) Tubs Only include tub, hoops, lugs, nuts, floor drain, chime joist set, two benches, and an instruction manual. Prices start at $1,160 for a 4 × 3-foot tub in redwood.

(2) Hot Tub Systems include the tub and a Spa Pack and are priced from $2,860.

(3) Convertible Hot Tub Systems include the tub, Spa Pack, and a vinyl liner with foam underlayment. Prices start at $3,255.

Staves for Almost Heaven hot tubs are milled with a ball and socket for a tight seal. Near their bottom, the staves are machine-notched to ensure a precise fit with the tongue-and-groove boards that make up the tub floor. Floor supports, called chime joists, are oversize for additional stability and to provide for ventilation. Hoops are made from ½-inch steel, vinyl-coated to prevent rusting. The tubs are available in 4-, 5-, 6-, and 8-foot diameters and 2-, 3-, and 4-foot depths.

The Spa Pack comes with a two-speed Jacuzzi pump, a Jacuzzi cartridge filter, and a Raypak 11KW heater, all preplumbed and mounted on a single redwood pallet. Also included in the Pack are four hydro-massage jets with stainless-steel escutcheons, two HydroAir suction fittings, a twenty-four-hour timer, 50 feet of flexible PVC hose, and a plumbing kit with all the ELs, TEEs, and nipples needed to complete the installation.

The Convertible closed-cell foam underliner gives the tub a soft, "padded" texture and increases the tub's insulative properties. The liner itself, in heavy-duty vinyl, is easy to clean and virtually puncture-proof. The tub

Almost Heaven barrel saunas are able to stand freely indoors or out. The sauna shown here measures 6 × 6 feet, holds four to six in comfort and, at $3,995, is the manufacturer's most popular model. *(Almost Heaven, Ltd.)*

can be used with or without the Convertible feature to suit seasonal needs. Since it installs over the tub's benches, with no water in the spaces beneath the benches, there's 25-percent less water to heat, which translates into additional energy savings.

Hot-tub accessories offered by Almost Heaven include a step kit ($120), a shelf kit ($60), a floating-weir tub skimmer ($35), a bubbler with air switch ($295), a Spa-Tron water purifier ($495), and a redwood tub cover (from $495) featuring tongue-and-groove construction and radial bracing.

We're also including Almost Heaven's barrel sauna in our coverage here. The sauna is able to stand freely indoors or out, but we think it would be a shame to hide one away when it so clearly belongs in a garden or other outdoor setting. The truly unique, 6-foot-diameter barrel sauna is constructed with kiln-dried, 2-inch clearheart redwood. With the front and back ends of the sauna delivered totally preassembled, on-site assembly takes only an hour or two. The door is prehung, using self-closing hinges, and the front and rear window glazing is tempered glass. Two slatted benches run the length of the sauna. A sauna light, a Tylö 6KW Sport Sauna Heater with rocks and built-in controls, a sauna cradle base, and an instruction manual complete the package.

The barrel sauna comes in 4-, 6-, and 8-foot lengths and is priced from $3,325.

Almost Heaven, Ltd.
Route 424
Renick, WV 24966
Phone: 304/497-3163
Product literature: catalog, $2

Snorkel Stove Wood-fired Hot Tubs

All you need is wood, water, and a match to enjoy a relaxing soak in one of the ingenious hot tubs turned out by the dedicated folks carrying on the art of cooperage at the Snorkel Stove Company in Seattle, Washington. Snorkel Stove hot tubs were originally created to bring the luxury of hot tubbing to remote Alaskan sites without electricity. But today they can be found in thousands of backyards, in both town and country locations, throughout North America and as far away as Ireland, Australia, and Saipan.

Before the introduction of the Snorkel Stove, there were only two ways to heat a hot tub—with electricity or gas, and a complicated system of motors and pumps to circulate the water past the heater and back to the tub.

Although created originally to bring the luxury of hot tubbing to remote areas without electricity, most Snorkel Stove hot tubs are purchased today just for natural, inexpensive tubbing. *(Snorkel Stove Company)*

Snorkel Stove's wood-fired hot tubs come in a number of sizes. Its popular 6 × 3-foot tub/Snorkel Stove package, in Western red cedar, costs $1,825. The tub can hold four to five comfortably. *(Snorkel Stove Company)*

Not only is the initial cost of the typical gas- or electric-heated hot tub nearly twice that of the Snorkel Stove System, but the monthly bill for the electricity or gas used can take a lot of the joy out of a leisurely soak.

With a Snorkel Stove, or its smaller sibling, the Scuba Stove, and an armload of wood, you can enjoy hot tubbing day or night, wherever you wish. No plumbing, pumps, or electricity are required. Except for the "snorkel" air intake, the chimney, and the feed door to the firebox, the stove remains submerged in the tub. As with any woodstove, a fire is prepared with paper, kindling, and some small logs. As the fire begins to burn, air is immediately drawn down the snorkel intake and a roaring draft feeds the fire. Adjusting the sliding door across the air intake regulates the rate of burn. When the water reaches tubbing temperature, the door can be closed, or left slightly open for a slow burn. If you overshoot the desired temperature, just add cold water to the tub.

The stoves, made from corrosion-resistant, heavy-duty marine-grade aluminum, are highly efficient at transferring heat from the fire to the water. A Snorkel Stove in a 5-foot-diameter tub holding 463 gallons will raise the water temperature 33°F. an hour, compared to 21°F. with an 85,000-BTU gas heater and

10°F. with a 12KW electric heater. A protective wooden fence separates tubbers from the hot stove. The stoves, which burn with little or no smoke, can also be purchased without the tub kit for installation as a new primary heat source in an existing hot tub.

The tubs themselves are easy to assemble. Most owners report that it takes about 3 or 4 hours to put one together. Old hands at the factory can assemble one in just 45 minutes. Complete instructions, covering assembly and operation, are included with each kit. There's also a toll-free customer support line.

Western red cedar or redwood tub-and-stove packages are priced from $1,420 for a 5-foot-diameter, 3-foot-high tub with a Scuba Stove. The 5 × 3 tub holds three adults comfortably. The most popular combination is the 6 × 3 tub with the Snorkel Stove. It costs $1,825 and can accommodate four to five adults.

Options include a Deluxe Additions package with a drain kit, bench set, shelf, and a two-part wood/aluminum cover, for $511. There's also a $48 Accessories Kit, which includes three 2-foot sections of galvanized stovepipe, thermometer, damper, ash scoop, and a bright yellow rubber ducky.

Snorkel Stove Company
108 Elliott Avenue West, P.O. Box 20068
Seattle, WA 98119
Phone: 1-800-962-6208
Product literature: free brochure

Sonoma Spas

Combining the look of the traditional hot tub with the easy maintenance and durability of acrylic, the Sonoma Round could make the all-wood tub obsolete. It's Sonoma Spas' most popular model. The 6-foot-diameter spa comes with four recessed hydrotherapy jets and a submerged "horseshoe" bench large enough to seat six adults.

The molded acrylic tub doesn't come as a kit. Neither does the redwood skirt that gives the spa its classic hot-tub look. Both come preassembled. However, that great-looking redwood spa surround *does* come as a kit, and that's why we've included Sonoma in our coverage of hot tubs and spas. That's not to dismiss the Sonoma spa line. Sonoma has been a leader in spa design and innovation since 1974, and is almost in a class by itself when it comes to high-quality spa shells and systems.

Sonoma's spa surround, erected in multiple levels, gives easy, graceful access to the tub and expands its capabilities. The components lend themselves to being "molded" for the most attractive as well as most useful arrangement for a particular setting. If installed over soft ground, paving tiles or other pier footings will be needed to provide solid, level support for the legs of the surround. Galvanized nails are supplied for the assembly, but screws would be preferable, with predrilling recommended to avoid any splitting of the wood.

The complete redwood deck kit, or surround, for the 35-inch-high wood-skirted tub costs approximately $700. Similar deck kits are available for the six other spas, which are either square or rectangular, in the Sonoma Spas line. List price for the 6-foot Sonoma

The Sonoma Round is a spa with the classic hot-tub look. The multiple-level redwood surround, which comes as a kit, gives easy, graceful access to the tub and expands its capabilities. *(Sonoma Spas)*

Round is $4,295. That price includes a form-fitting, thermal-insulated cover for the spa.

Sonoma Spas
5845 193rd Avenue
Rochester, WA 98579
Phone: 1-800-772-4762

Houses

Unlike most other kit items, no two kit houses are necessarily alike. Most major manufacturers of kit houses have turned to computer-aided design (CAD), and even computer-aided manufacturing (CAM), which allows for easy customizing of any standard plan the manufacturer might have in inventory. This means kit-house buyers can get a specially designed package that fulfills their wish list without having to pay an architect's high fees. In a structure made largely of wood, there is virtually no limit today to the choices available to the kit-house buyer—whether that's for post-and-beam or panelized construction, a traditional or a contemporary design, a log home or a geodesic dome.

Building with a kit almost always represents significant savings over a house that is "stick-built." With the use of both precut and factory-assembled prenumbered components, the shell of the house can be erected and made weathertight in far less time than would be required by the same crew when nearly everything has to be cut and pieced together from raw materials brought to the building site. Even at most owner-builder schools, the recommendation to students now is to build their homes with a kit rather than from scratch.

A course in the fundamentals of house-building *is* useful but not a prerequisite for participating in the construction of one's own home. With most kit houses of any size, the wise buyer leaves the erection of the shell to the pros. The buyer may elect to serve as the general contractor though, and even more likely will take on some of the finishing work. According to the folks at Miles Homes, which deals only with those who are prepared to build without a lot of outside help, one out of every three new homes built in America this year will have some degree of hands-on owner involvement.

Before you approach a kit-house manufacturer with checkbook in hand, though, pay a visit to your local zoning board. You could run into a few roadblocks, especially if you are planning to go the d-i-y route. Under local codes, you may not be permitted to build the kind of house you have your heart set on, even on land you already own. You may be restricted as to the extent of your participation in the actual construction, or even the overseeing of the construction; you may not be permitted to build without a licensed contractor. As an inexperienced builder, you may be limited to a house of one-story design. In many built-up communities, you'll also likely find that local building codes require the engagement of licensed plumbers and electricians to handle the installation of the so-called mechanical systems.

In short, don't count on reaping savings through "sweat equity" until you know for certain what's allowed under local zoning and building codes. And don't think for a minute that constructing a house from a kit is all that quick and easy, even though it is possible, with a competent crew, to complete the shell of many kit houses in a matter of days. The kit often represents little more than the shell of the house, and, unless you're building with logs or closed-wall panels, even at this dried-in stage you're likely no more than a third of the way home. Installation of the mechanical systems, insulating and closing the interior walls, hanging interior doors and wall cabinets, plastering, trimming-out, painting, tiling, and floor sanding and finishing can take just as much time with a kit house as it does with stick-built construction.

It's been said that constructing your own home, even from a kit, is not unlike eating an entire elephant. It's bite by bite by bite until the job gets done. To effect additional savings then, you're going to need time and plenty of it—commitment, too. If you have a regular job, that's going to mean working during your vacation, working most holidays and weekends, and, if the site is within an easy drive of your present residence, working more than a few evenings.

Don't plan on camping out in the house in order to maximize your time, though. Any occupancy of the house, other than during working hours, will likely be barred until you have been issued a certificate of occupancy by a building inspector, who will have been checking the work for code observance from time to time. We know of kit builders who were denied occupancy until the last window sash had been painted and the last smoke detector had been wired into the electrical system.

Acting as your own general contractor can be just as demanding of your time. You are going to have to be at the site, or no more than a phone call away, nearly every day until each job is completed. If your regular employment doesn't permit this close participation, your spouse or another family member might take on some of the responsibility. But one or both of you are going to have to secure building permits, solicit bids, hire subcontractors and others, purchase finishing materials, arrange delivery and work schedules, adjust to bad weather, and be prepared for the other delays that are time-consuming and contribute to cost overruns. But don't ever despair. There is little in life that is more satisfying than having built one's own home.

You'll likely get a taste of contracting early on, when you have to arrange for construction of a foundation for your house. Specifications for it will have been included in the plans, but that's the extent of the kit manufacturer's participation here. It's another job that's best left to the professionals, since, if the foundation is even a little bit off, your preengineered house isn't going to fit properly. An experienced crew, however, working with an earth mover and specially designed wall forms to contain the poured concrete, can make quick and exact work of this stage of construction.

You might gain some experience, too, in arranging for the site prep, which will have to include an access road of sufficient width and bottom to allow delivery by one or more flatbed trailers carrying up to 40,000 pounds each of precut logs or lumber and the other components that will eventually make up your house.

In 1978, when we wrote *The Complete Kit House Catalog,* the first book to give coverage to the full range of precut and panelized house sales packages, kit-delivery charges, at \$1 to \$1.50 per mile per trailerload, were an expense that had to be given serious consideration. In relation to the then cost of most precut packages, it didn't make much sense to deal with a manufacturer or distributor located more than a few hundred miles from the building site. But today, while you might pay $2 per mile per trailerload, the cost of the kit itself is likely to be several times what it would have been in 1978. As a consequence, the market area for many of those same manufacturers, selling through a network of dealers or distributors, has gone from regional to nationwide. Indeed, many are now selling their kits worldwide, since, with containerization, delivery to distant ports by ship is often no more expensive than it would be for a long-haul delivery within the contiguous forty-eight states.

There's no easy formula for determining the final cost of any kit-built house. A lot depends on the extent of your participation and your success in hiring others to do those things that you have neither the time nor the talent to take on yourself. Act as your own general contractor, though, and you might save between 10 and 15 percent of what a fully-contracted house would have cost. Do your own plumbing, wiring, and finishing, and you likely could save another 10 to 15 percent.

You can't go by kit price alone, since the kit for a log home might include no more than the precut wall logs and perhaps windows and doors, while the kit for a house of traditional design might come complete to coat hooks and kitchen cabinets. Excluding land, but factoring in site prep and all other expenses, a log home or a dome might end up costing three to four times what you will have paid for the kit. On the other hand, a house of contemporary or traditional design, built from a kit complete enough to be worth its $60,000 price, might end up having cost no more than twice the kit price.

But let's take a look at the best in a wide range of kit houses. There just might be a kit here that could put the home of your dreams within reach.

Bow House

From time to time, New York's Museum of Modern Art mounts a show of manufactured products, both old and new, chosen for the excellence of their design. If MOMA has room for one, the selection committee would

The distinctive curved roofline of the Bow House does more than merely please the eye. It provides added headroom where badly needed and shrugs off rain and heavy snow loads. The model shown here is the three-quarter Cape, with an ell addition. *(Bow House, Inc.)*

do well to consider a Bow House for inclusion in this select company.

If the buyer follows Bow House's detailed building manual and uses its unique Architectural Package of reproduction building materials, every Bow House has that museum-quality potential worthy of the restorations of Deerfield, Sturbridge, and Williamsburg.

The distinctive Bow House roof is the rebirth of an early New England coastal practice of bending roof beams to please the eye of the seafaring homeowners of the time and to withstand the fury of Nor'east gales. Bow House produces curved, laminated rafters to create a roof that is even stronger than the ones on those pre-Revolution homes, some of which are still standing. The earliest extant bowed-roof Cape was built in 1678!

There are four basic Bow House designs, ranging from the 1,182-square-foot quarter Cape, through the half Cape and the three-quarter Cape, to the 2,563-square-foot full Cape. The full Cape measures 42 × 28 feet and, since the curved roofline provides extra headroom where it's badly needed, is large enough to allow for four upstairs bedrooms and two baths. The 26 × 20-foot quarter Cape is more suited for use as a guest cottage or single-person residence.

Those who decide to build a Bow House will find themselves working as closely with the Bow House design staff as they might with their own architect. With this exception: design costs are all part of the Bow House

The Bow House half Cape is a faithful reproduction of the curved-roof houses common to New England coastal villages three-hundred years ago. The Architectural Package for this two- or three-bedroom house shown with a gable-end addition costs about $45,000. *(Bow House, Inc.)*

An owner-builder, working as general contractor and doing much of the finishing work, might bring in this authentic-to-the-period, four-bedroom full Cape, with gable-end extension and 24 × 24-foot saltbox garage, for under $140,000. *(Bow House, Inc.)*

overhead. Every plan is extensively customized, if required, to meet the owner's needs, just so long as those exteriors of the house most visible to the public respect the Bow House concept. Any modern deviations, such as a sun room, are usually positioned at the rear of the house, as are skylights installed to provide more natural light for the upstairs.

Unless you're going to build the house yourself, it's up to you as the buyer to find a local builder who will construct the house for you. In soliciting bids, the Bow House buyer must take into consideration those building materials that are not part of the Architectural Package. Common framing materials, whether for post-and-beam or more conventional studwall construction, are not included with the components that will be delivered by Bow House. Plywood sheathing, 2×4s, insulation, and drywall can be purchased at local lumberyards and building-supply centers. The Architectural Package can then be delivered on a single truck, which halves the transportation costs.

It's the Architectural Package that *makes* this house. It includes a long list of items, from the laminated roof rafters, handmade exterior doors, and historically correct divided-lite windows to beveled red cedar clapboard siding, exterior pine trim, and white cedar roof shingles. Also included are handmade interior doors, stair parts, 12-inch-wide floorboards, fireplace mantel kits, and interior trim.

The Bow House building manual is one of the most useful we've seen and incorporates more than twenty years of suggestions from Bow House builders and owners. It also provides pages of tips on how to make your home true to the colonial period, if that's your wish.

Bow Houses are not confined to New England. They've been built from Maine to California, by expatriate New Englanders as well as others who have a sense of history and are

attracted by the aesthetic appeal of the Bow House's proportions and clean lines.

Package prices for Bow Houses run from about $30,000 to $50,000. With the additional materials costs and excluding any input by the owner, the completed house price, including allowance for the foundation, appliances, kitchen cabinets, and the like, might run from three to four times the price of the Architectural Package. Two- and three-car saltbox garage packages are also available, at from about $5,000 to $6,000.

Bow House is also responsible for Bow-Bends pool and garden structures (see page 95).

Bow House, Inc.
Bolton, MA 01740
Phone: 508/779-6464
Market: national
Product literature: Bow House brochure, $7.50

Lindal Cedar Homes

Lindal is the world's largest manufacturer of custom cedar homes and very likely the best-known name in the kit-house industry. In the home-manufacturing business since 1945, Lindal has come a long way from its post–World War II A-frames and other budget precuts.

Just take a look at any of the Lindal homes pictured here.

The basic ingredient of every Lindal home is kiln-dried Western red cedar. Why cedar? It's one of nature's finest building materials. It's a tight, low-maintenance wood, with good insulting qualities and a natural resistance to insects, vermin, moisture, and decay. Pound for pound, it's stronger than steel. It's also a beautiful, durable wood that, like a fine wine, only improves with age.

As one of the first kit manufacturers to have turned to computer-aided design and cutting, Lindal can rightfully claim that there are as many different Lindal homes today as there are Lindal homeowners. Whether your interest is in the traditional, the contemporary, or a modern original, Lindal is fully prepared to meet your needs and preferences.

The Lindal network of over 350 dealers and distributors throughout North America, and in Europe, Japan, and Australia, is the largest in the industry. Using a Lindal Home Planning Kit, the prospective Lindal homeowners are encouraged to put their dreams down on paper. Then, in consultation with a Lindal representative, the customers' concepts are refined. A plan elevation and cost estimate are then developed by the Lindal organization for the customers' consideration—completely free of charge and with no obligation.

Prospective homeowners always have the option of choosing from one of the proven designs, comprising thirty separate series and 101 models, contained in Lindal's spectacular, 240-page, four-color idea book. Since Lindal offers the option of either post-and-beam framing or solid-timber-wall construction, both of which support upper levels and the roof system with no need for interior load-bearing walls, these, too, can easily be customized. All it takes is a few minutes at a computer terminal to move a bedroom wall, add a cantilevered deck, knock out a closet, or expand a bathroom to accommodate a spa.

Dramatized by a two-story sun room that doubles as a stunning formal entry, the three-bedroom, two-bath Sunburst, from Lindal's Designer series, is a modern classic. Price for the kit, which is as nearly complete as they come, is approximately $85,000. *(Lindal Cedar Homes)*

(*above*) If you need any convincing of Lindal's ability to custom-create a home for its site, the well-named Paradise Found, with driftwood siding and Pacific-blue roof tiles, was owner-built in Kailua Kona, Hawaii, from a package costing under $100,000. *(Lindal Cedar Homes)*

(*below*) The appeal of Lindal's popular Summit series begins with its distinctive roofline and panoramic prow front. The unit shown here was built using the Justus system solid-timber-wall option at a kit price of under $50,000. The decking is an extra. *(Lindal Cedar Homes)*

A pleasing, practical design that uses space efficiently and lends itself to the zoning of activities, Lindal's four-bedroom, three-bath Burgundy, in the Colonial series, has a kit price of under $70,000. This works out to around $30 per square foot. *(Lindal Cedar Homes)*

The owners of this Colonial series Pilgrim added an intersecting gambrel and a one-story wing to what began as a 1,000-square-foot core. The three bedrooms are zoned upstairs and to the rear of this 2,313-square-foot house. Kit price is under $65,000. *(Lindal Cedar Homes)*

The beauty of post-and-beam, of course, is that it opens up the interior to all kinds of design possibilities, not the least of which are planked cathedral ceilings and exposed beams. Unlike conventional studwall construction, in which both interior and exterior walls bear the weight of the upper levels and the roof, sturdy, widely spaced posts and glue-laminated beams shoulder those loads in a Lindal home.

The solid-timber-wall option was introduced by Lindal in 1983, when it acquired Justus Homes. The Justus system evolved from early log construction, but improved on it in many ways. Precision-milled 4 × 8-inch solid cedar timbers, double tongued and grooved for an airtight fit, are locked together horizontally to form the walls. Where walls intersect, dovetail joints prevent movement and slippage, avoiding the energy-loss and settling problems of many log homes. The smooth-faced timbers not only provide exterior "siding" but insulation and an inside finished wall as well. This is known as single-wall construction and could be a good choice for those who want to build their own homes and achieve significant labor savings. As to the insulating qualities of the Justus wall system, the thermal mass generally performs equal to or better than cavity walls with R value of 19.

Kit prices for the houses in the Lindal plan book range from just under $20,000 to around $130,000. Add approximately 15 percent to those prices if you want the Justus timber-wall package. Lindal also offers clapboard and round-log options. Garage packages are also available. Prices include all materials necessary to make a weathertight home. All doors, floor systems, stair assemblies, insulation, trim, nails, locks, and caulking are part of the package.

Lindal Cedar Homes
4300 South 104th Place, P.O. Box 24426
Seattle, WA 98124
Phone: 1-800-426-0536
Market: worldwide
Product literature: Lindal Planbook, $18
Video: *How a Dream Comes to Life,* $19.95

Northern Homes

Since 1945, Northern Homes has built or supplied the kits for more than thirty thousand quality homes, most of which have been based on traditional or colonial designs. Named for such Revolutionary-era figures as Thomas Jefferson, Ethan Allen, Sam Adams, Betsy Ross, and John Hancock, more than a few of these houses are so faithful to the period they look as if they belong in the National Register of Historic Homes.

While it has built an enviable reputation along the eastern seaboard—with a network of dealerships extending from Maine to Ohio to North Carolina—Northern Homes' market area also includes California, Japan, Spain, and the Mideast. You're not likely to find an Ethan Allen or a Betsy Ross in any of the desert kingdoms, but then a large proportion of the houses built or packaged by Northern Homes are one of a kind, or adapted from one of the many stock plans in an inventory that today includes not only New England traditional but Georgian, Greek revival, Tudor, and American contemporary.

True do-it-yourselfers represent less than 10 percent of Northern Homes' customers. Another 30 percent may do some of the finishing work themselves, after the shell has been erected. It's not that the frame of the house is all that complicated, but many of

(*above*) A saltbox design faithful in both scale and detail, the 3,133-square-foot Ethan Allen is an excellent example of Northern Homes ability to replicate period houses. The plan allows for four bedrooms, plus a guest room over the garage. Kit price is approximately $112,000. *(Northern Homes)*

(*below*) The traditionally styled, 2,680-square-foot Hobbs Mill, a recent addition to the Northern Homes Design Collection, has gained immediate popularity. Window treatment is a key element in the Hobbs Mill's appeal. Kit price for this three-bedroom, two-and-a-half-bath house is under $80,000. *(Northern Homes)*

these homes are huge, multilevel affairs. This kit manufacturer presents its home designs in two separate catalogs: *The Design Collection* and *The Founders' Guild Collection*. The latter catalog features thirty-nine houses, with floor areas ranging from 3,400 to 9,600 square feet. A couple of the houses in this collection wouldn't look out of place in France's chateau country! Median *kit* price for a home from this collection is $162,440, FOB, Hudson Falls, New York.

Factory-assembled wall panels, with preframed door and window openings, can translate into significant reductions in the time and labor costs of erecting the dried-in shell on any Northern Homes design, whether stock or one-of-a-kind custom. *(Northern Homes)*

Northern Homes does simplify things a bit, though, since the walls of its houses are made up from sections preassembled in the factory. These are of open-wall design and represent only the studwall and exterior sheathing. Designed for immediate assembly, wall panels and the roof-system framing are delivered to the site once the first-stage flooring is in place. Where possible, to minimize exposure on-site to theft, vandalism, and the vagaries of weather, Northern Homes stages its deliveries.

If the houses in *The Founders' Guild Collection* are beyond your skills and means, you *might* find what you're looking for in *The Design Collection*. The houses here, in a broad variety of styles, have a median size of 2,350 square feet and a median kit price of about $74,000. There are only three kit packages priced under $50,000.

It would be hard to find fault with any of the Northern Homes designs, though. Product improvement at Northern Homes is a continual process, and only those designs that have remained functional and attractive over a long period of time are retained in inventory.

Another plus here: Northern Homes dealers or sales representatives are trained to manage the intricacies of securing permits, meeting code requirements, and analyzing site preparation. Their readiness to help with many of the jobs that would normally fall to the contractor is just one of the reasons that not a few buyers of these homes elect to act as their own contractors.

The package you receive from Northern Homes represents everything needed to construct a trimmed-out cedar-sided and shingle-roofed shell, plus an interior package that, with the exception of kitchen cabinets, includes interior doors, partitioning, stairs, trim, and just about anything else you might need that's

A Northern Homes three- or four-bedroom contemporary design, the 2,839-square-foot Pine Valley includes a three-car garage and features an entry foyer and living room that open dramatically to a vaulted second-floor ceiling. Kit price is approximately $75,000. *(Northern Homes)*

constructed with wood. Insulation is priced separately as an optional item. Northern Homes millwork capabilities also make it able to provide many custom wood products.

Northern Homes, Inc.
51 Glenwood Avenue
Queensbury, NY 12804
Phone: 518/798-6007
Market: Eastern states, California, Japan
Product literature: *The Design Collection,* $20; *The Founders' Guild Collection,* $20. Both prices include a copy of *The Consumer's Home Building Guide.*

Timberpeg

Timber framing, one of the oldest forms of building, was rediscovered by house builders in the early 1970s. Used for centuries in the construction of barns, which are hardly known for their energy efficiency, it does lend itself to the building of energy-efficient houses.

Closely related to post-and-beam construction, the posts and beams in a true-to-the-past timber frame are fitted together by mortise-and-tenon joinery, with the intersecting timbers secured by strong pegs driven through predrilled holes. Additionally, all of the materials that make up the exterior walls are applied to the outer face of the posts. This leaves the sturdy frame dramatically exposed within the home.

The energy saving comes about when a thermally efficient envelope is wrapped around the perimeter of the frame. This might be made up of stress-skin panels (a preassembled sandwich of rigid insulation and sheathing), or built up with a sequence of finishing, insu-

This dramatically situated Timberpeg achieved everything the owners were hoping for: spacious rooms, cathedral ceilings, and the warmth and beauty of a timber-framed interior. The home sits on a hillside—its balcony and windows overlook the town and valley below. (Architect, Robert J. Daly, AIA; Photographer: Ross Chapple).

(right) The living room of this New England home features a full height, cathedral ceiling, semicircular balcony and native stone chimney. The exposed pine framing and expansive glass create a room which is casual and filled with light, while the Timberpeg wood frame is both elegant and functional. Using the structural framing and masonry as decorative elements, this home shows some of the creative possibilities of Timberpeg framing. (Photographer: Ross Chapple).

A Typical Timberpeg, Layer By Layer

1. Conventional framing at first floor (supplied by others).
2. Timberpeg pre-cut mortise and tenon frame.
3. 1"x8" tongue and groove pine.
4. 6-mil polyetheylene vapor barrier (supplied by others) applied directly over 1"x8".
5. Polyisocyanurate foam insulation board.
6. 1"x4" strapping.
7. 1/2"x6" cedar clapboards.
8. Cedar trim.
9. 1"x4" shake or shingle ribs.
10. 30-pound felt interlayment (supplied by others) between cedar shake layers.
11. Western red cedar shakes or shingles.
12. 1"x8"cedar ridge cap.
13. Tyvek (or equal) air barrier (supplied by others).
14. Andersen® Perma-Shield Low E windows—style per plan.

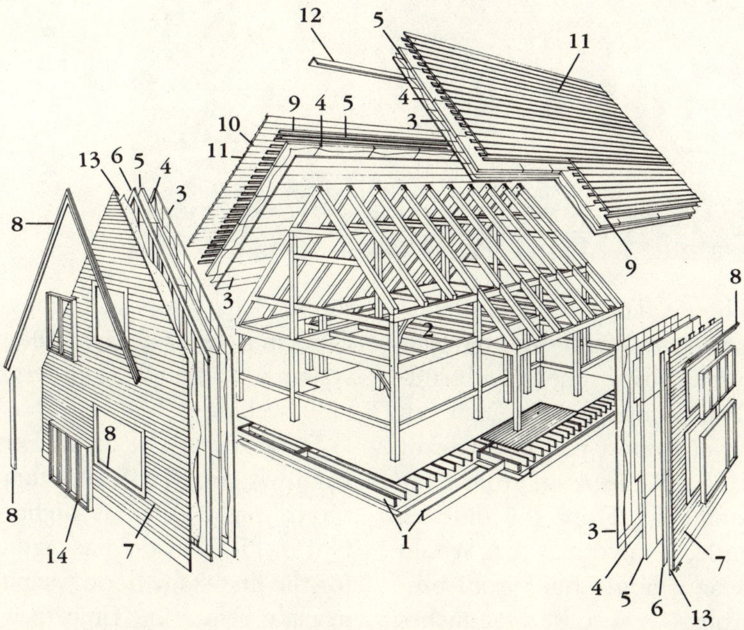

lating, and sheathing materials. The trade-offs here are that while stress-skin panels can cut down on the time it takes to enclose the house, they can be too heavy to handle without mechanical aid and they're a more expensive option than sequential, outer-wall construction.

Timberpeg, the kit-house manufacturer we've elected to focus on here, was one of the first to revive this centuries-old method of construction. They're now a national company and have manufacturing facilities in New Hampshire, North Carolina, Colorado, and Nevada. As a prospective Timberpeg owner, you would deal with a Timberpeg Independent Representative, many of whom are well-respected builders, engineers, designers, or architects in their own right. They not only know how to build a Timberpeg home but can help you through all the other complexities of getting that home from conception to the finished stage.

Timberpeg prides itself on being "the artisans of post and beam." Glancing through the more than 100 pages of a catalog filled with in-depth ideas and advice, insightful articles and how-to pieces, colorful photographs, and over fifty original designs for eye-pleasing Timberpeg homes in all sizes, floor plans, and styles provides confirmation of the claim.

Since the frame of a Timberpeg is self-supporting, interior walls can be placed any-

This is not the face of a saltbox that one normally expects to see, but with Timberpeg the idea is to take those elements you like from any number of its custom and stock designs and let Timberpeg make the home of your dreams come true for you. *(Timberpeg)*

where within the house. This, of course, lends itself to wide-open spaces and soaring ceilings—and, with the exposed timbers, a blending of traditional and contemporary. You can achieve just about any effect you might want within the walls of a Timberpeg, from cozy rustic to truly dramatic. From the outside, though, there's little, if anything, that would tip this off as being a house built from timbers measuring from 4 × 4 to 8 × 12 inches and up to 16 feet long!

Timberpeg packages include precut-and-planed Douglas fir or eastern white pine posts, girders, rafters, ties, joists, and angle braces, with mortise-and-tenon joinery and the oak pegs needed to secure them. The standard wall system includes 1 × 8-inch tongue-and-groove pine, milled for a V-joint, finished interior wall; 4 × 8-foot panels of rigid-foam insulation; 1 × 4-inch strapping, to create an airspace that gives additional insulating value to the total wall system; and ½ × 6-inch Western red cedar clapboards for the exterior finish.

The roof system is much like the wall construction, with pine decking, insulation board, roofing felt, 1 × 4-inch shingle ribs, and then hand-split cedar shakes. The upper-level floor system includes joists, girders, and 2 × 8-inch pine decking.

These materials, as well as all doors and windows, are of a quality that other manufacturers might offer as higher-priced options. Not included in the package are the materials for the first-stage floor system, which must be in place before the Timberpeg package arrives.

Timberpeg preengineered package prices range from around $35,000, for a 1,000-square-foot, one-bedroom, one-bath Carefree that lends itself to later expansion, to $158,000 for the 9,040-square-foot Grosse Point, which, in addition to its three primary bedrooms, has a secondary three-bedroom wing that would be perfect for guests, in-laws, or household help. Estimated completed Timberpeg costs, fully contracted, range from $70,000 for the Carefree to $390,000 for the Grosse Point. There are ten series of Timberpegs, all but two of which include models that could be completed for $150,000 or less.

How much could you save by doing some of the work yourself? Since erection of the timber frame calls for familiarity with the process and sometimes the use of a crane, the frame is almost always contractor-installed. But then a contractor and an experienced crew can put the frame of most Timberpeg homes together in anywhere from 2 to 5 days. On your own, working with ropes and pulleys, or even a few husky buddies, it could take many weeks. Mounting stress-skin panels, especially above the first course, could also be beyond your capabilities.

But go with the standard wall system, and it shouldn't be that great a challenge. Working with a Timberpeg representative, acting as your own contractor, and doing as much as you can once the frame is up, you might save up to 25 percent of what a fully contracted house would have cost.

Timberpeg specializes in custom homes, and if your land has a great view, they can develop a package that will take full advantage of it, as does this striking, oceanfront home, designed along an elongated axis and capped by a rooftop widow's walk. *(Timberpeg)*

Timberpeg East, Inc.
Box 1500
Claremont, NH 03743
Phone: 600/542-7762

Timberpeg South, Inc.
Box 880
Fletcher, NC 28732
Phone: 704/684-1722

Timberpeg West, Inc.
Box 270988
Fort Collins, CO 80527
Phone: 303/221-3355

Timberpeg Pacific, Inc.
Box 70123
Reno, NV 89570
Phone: 702/826-4447

Market: United States and Canada
Product literature: Timberpeg catalog, $15

Topsider Homes

It's quite fitting that the original Topsider home, The Treehouse, sits on a pedestal. Introduced in 1968, it's been one of the most successful of all preengineered houses. Thousands of units have been built worldwide, from Disney World in Florida to Hakone, Japan.

The building system that has made The Treehouse so popular has spawned a large and growing family of Topsider homes, including larger and smaller versions of the pedestal-topping Treehouse. In addition to these siblings, there are patio models, such as The Trailsider and The Campsider, which are built without a pedestal. There also are two-story Topsiders, such as the popular Treetopper, and The Slopesider, a two-story adaptation of The Treehouse.

By whatever name, all Topsiders are eight-sided, with lots of floor-to-ceiling insulated glass, for panoramic viewing; vaulted, exposed-beam ceilings, and no interior load-bearing walls. The Topsider building system permits total floor-plan flexibility and adapts to virtually any climate, setting, or terrain.

The basic shell package for The Treehouse allows for 900 to 1,000 square feet of usable floor area. If you require more space, you could move up to the 1,250-square-foot Shoresider or the 1,405-square-foot Viewsider, which are larger versions of The Treehouse. You could also go to one of the larger two-story variations, or cluster two or more units, which makes for easy zoning and has become a popular choice with many who are building a primary residence rather than a vacation home. Other options would include enclosing lower-level space with a pedestal model or adding on a stick-built addition to a patio model.

Most Topsider Homes customers take advantage of the opportunity to work with its Architectural Services Department in putting

The eight-sided Treehouse sits serenely atop a 10-foot pedestal, and rightfully so. It is one of the most popular houses ever offered in kit form. The building system can be adapted to accommodate virtually any terrain, climate, or size requirement. *(Topsider Homes)*

Blending into the natural landscape as if they grew out of the ground, Topsider's pedestal homes are grown-up answers to kids' dreams. Framed with floor-to-ceiling insulating glass, they take better advantage of most sites than do a mere four walls. *(Topsider Homes)*

together a package that fully meets not only their preferences but seismic, tropical, and snow-country building requirements. More than just a space designed for living, Topsider homes are a blending of form and function to meet every need, no matter where you plan to build.

The floor, walls, and roof of a Topsider home, the framing for which is essentially post-and-beam, are assembled from preengineered panels of closed-wall design—with insulation sandwiched between interior and exterior finished surfaces for the walls, and a roof that's ready for its final cover. These homes go together quickly, even when set atop that pedestal. If you choose, Topsider will rent you a set of forms for casting the reinforced-concrete base. A 20-minute video is also available, showing construction of a pedestal model *From the Ground Up*.

Shell packages for Topsider homes include all structural elements (except concrete and reinforcing steel for pedestal or foundation), insulated roof panels, roof trusses, roofing

The Skisider is Topsider's most compact pedestal home. Ideal for difficult sites, here it serves as a refuge high among the slopes. With up to 675 square feet of floor area, it also makes a neat one-bedroom home or efficiency. Package price is $18,950. *(Topsider Homes)*

More and more Topsider customers who are building homes to be used as full-time residences find that clustering not only meets the needs for more space but makes for easy zoning—with the family and children's rooms separated from adult bedquarters. *(Topsider Homes)*

(*below*) Once the steel-reinforced pedestal and cantilevered floor system are in place, the prefabricated wall and roof sections of The Treehouse, or one of its larger or smaller siblings, go together quickly and easily, frequently in just days. *(Topsider Homes)*

felt, wall frames, windows, doors, and corner panels. Pedestal and two-story models include floor panels, floor trusses, and, where applicable, pedestal basement-closure materials. Specially milled exterior and interior trim, as well as all fasteners, are part of the package.

If you prefer something more down-to-earth, the basic Topsider can be constructed on the flat, or even on sloping ground, as shown here. Once the shell is "weathered-in," conventional construction techniques are employed to complete the interior. *(Topsider Homes)*

Since there are no interior load-bearing walls, Topsider owners have complete freedom when it comes to floor-planning. Acute angles are avoided by installing divider walls perpendicular to exterior walls rather than bringing them out from corners. *(Topsider Homes)*

Topsider offers an interior package as an option to its domestic customers. This includes two preplumbed modular bathrooms; a complete kitchen with prefinished cabinetry; doors, and prewired panelized partitioning, to create two bedrooms, closets, and living and dining room areas. Cost of the option for a standard pedestal model is approximately $8,500.

Domestically, most Topsider interiors are stick-built, using locally available materials. No special skills or techniques are required, other than those used in conventional house construction. The interior package is meant for builders abroad, and here Topsider Homes can put together a package that includes everything needed for a "turn-key" home, down to linen and silverware. This has come about because a lot of Topsider homes are built by resort-area developers.

It's hard to put a cost on a Topsider. There are many variables, including the model purchased, options, modifications, where it is erected, and by whom. Construction and finishing costs will typically add from 50 to 100 percent to the home package price—for anywhere from $40 to $70 per square foot finished.

Basic-shell package prices range from $12,990 for a 575-square-foot Trailsider, Topsider's most popular patio home, to $33,500 for the 1,800-square-foot Treetopper, its most popular two-story model. Price of the basic Treehouse is $27,350.

T.H. Manufacturing, Inc.
P.O. Box 849
Yadkinville, NC 27055
Phone: 919/679-8846
Market: worldwide
Product literature: information kit, $10
Video: *From the Ground Up,* $30 refundable deposit

Real Log Homes

When we first wrote about the return of the log home back in the mid-1970s, we were hard-pressed to come up with the names of ten established log-home manufacturers. Today, there are upwards of 450 log-home producers, and the industry, which has gone international, represents a \$2-billion-a-year business.

While it can come down to a hard choice when selecting a company from which to order the makings for a log home, most of those 450 companies won't even be in the picture. A majority represent small, local operations. About 85 percent of all log homes today come from some fifty manufacturers, and our choice here, Real Log Homes, gets a good chunk of that business. With manufacturing facilities in Vermont, North Carolina, Montana, and Arizona, and a network of more than 200 Independent Representatives in the United States, Canada, Asia, and Europe, Real Log Homes is the nation's largest precut log-home producer. It is also the most progressive.

As the first log-home manufacturer to have national distribution, the first to be approved by all three National Code Agencies and the Japanese Building Code Authority, introducer of the patented Interlock log-joining system,

At home in any setting, the distinctively styled Cascade offers a fresh look at log-home living in the 1990s. A 21 × 27-foot great room with a cathedral ceiling and a loft walkway are but two of the unexpected pleasures in this three-bedroom contemporary. *(Real Log Homes)*

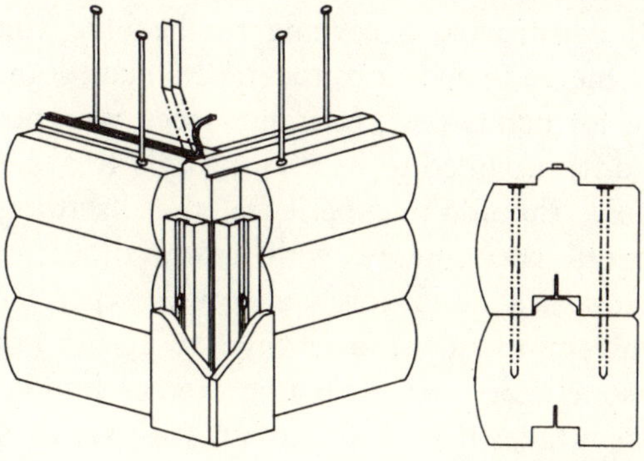

V-Groove Flat Interior

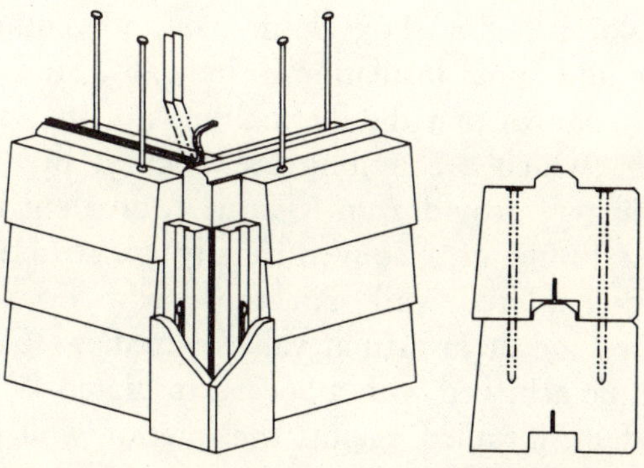

Clapboard

Real Log Homes options in log styles and corner treatments are far removed from natural log shapes and traditional notched-end construction. It's these and other such advances that are bringing log homes out of the woods and into the suburbs. *(Real Log Homes)*

and with more than seventeen thousand homes built since 1963, this is a company from which business majors might take lessons.

We don't intend to make the case for building a log home here. We will let the success of the fastest-growing segment of the home-building industry speak for itself. Make no mistake, though, log homes today are a far cry from early log cabins. They've not only come out of the woods and into the suburbs, but better than 90 percent of them are being built to serve as primary residences.

In its plan book, Real Log Homes offers as wide a variety of home styles, models, and floor plans as any company in the log-home business. From rustic cabins to expansive solar contemporaries, from wraparound porches and cathedral ceilings to prow fronts and gable ends, there's very little that can't or hasn't been done with a Real Log Home. Inside, it's often a blend of log walls, sheetrock, wallpapered rooms, and tongue-and-groove paneling.

LOG WALL ASSEMBLY DETAIL

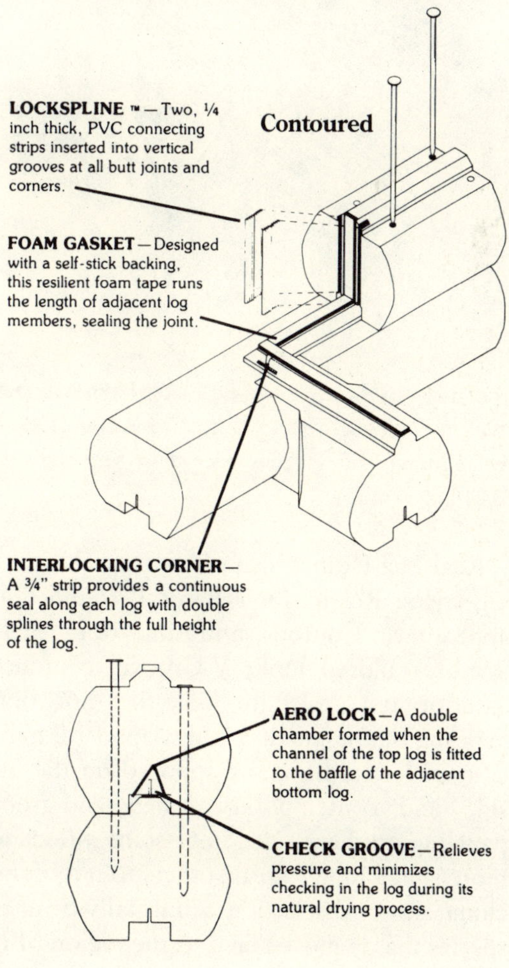

Since the introduction of computer-aided design, virtually every Real Log Home has been a custom job. In consultation with their Independent Representative during the planning stages, customers are encouraged to request any design changes that might make their Real Log Home a more perfect place for them.

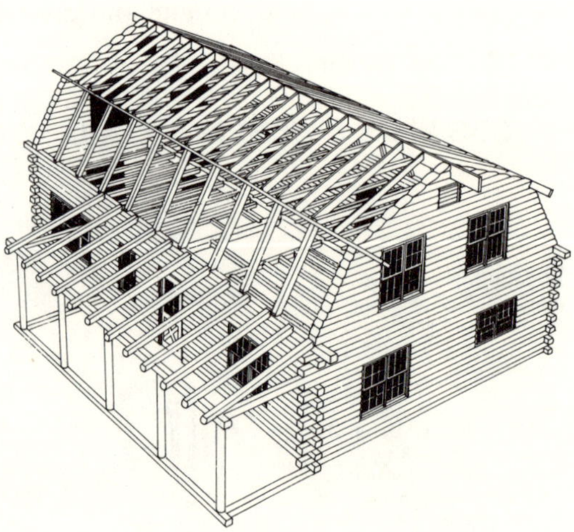

The materials package for Real Log Homes popular gambrel-roofed Plymouth B costs $30,600 FOB the manufacturer and includes everything you see here. (Real Log Homes)

Real Log Homes also offers four choices of log styles: Rustic, for a natural, rounded-log appearance; Contour, uniformly rounded for a more tailored look; V-Groove, contoured exterior and milled to look like horizontal paneling on the interior, and Clapboard, milled to pass for traditional clapboard on the outside and contemporary tongue-and-groove paneling on the inside. Since some woods are more suited to dry areas, others to hot, humid climates, wood selection will usually be of the species that is most suited to the region of the country where you intend to build.

If constructed according to the blueprints and the 162-page Construction Manual (looseleaf, for handy page reference when you most need it) provided to its customers, a Real Log Home shouldn't experience the "settling" horrors encountered with many other log homes. All solid log walls will experience some contraction and settling, no matter how seasoned the logs. Real Log Homes building system takes this into account, with vertically installed Lockspline and foam seals that allow the logs to settle without compromising the airtight seal around windows and doors.

Nor is a Real Log Home likely to require any additional insulation in its walls. Studies have shown that the thermal mass of its wall logs, which are among the thickest in the industry, provides an R-value equivalent to that found in a conventionally constructed and insulated wall. For those who do require it, an insulation value exceeding R-25 can be achieved with a Special Insulated Wall System, installed against the interior wall of a Real Log Home.

Almost all Real Log Homes represent single-wall construction: outside finished wall, insulation, and inside finished wall all in one. This building system has made log homes a popular choice with owner-builders who want to do as much of the work as they can themselves. It's not easy work, though. It can take muscle to lift the longer logs and to swing a spike-driving sledge.

What do you get in a typical Real Log Homes package? Not as much as you might in a package for a conventional precut home. You do get all the logs, precut and numbered, for the walls; prehung doors and windows; girder and log joists; roof rafters and collar ties; porch posts, sills, plates, and rafters; log gable ends; 10-inch spikes, Locksplines, and foam gasket tape. You get everything needed

The 2,834-square-foot Woodstock B incorporates all of the features of the Plymouth, with a two-story gable wing added for extra living space. The plan allows for four bedrooms, plenty of closets, a large living room, plus a separate family room. *(Real Log Homes)*

Log-home living can be as elegant as you want to make it. From cathedral ceilings to a blend of log walls, sheetrock, wallpapered rooms, and tongue-and-groove paneling, there's very little that can't or hasn't been done with a log-home interior. *(Real Log Homes)*

for a weathertight shell except for the roof covering. As with most log-home manufacturers, the customer, with the aid of the Independent Representative, is expected to acquire the more common building materials locally.

Log package prices for Real Log Homes run from about $15 to $25 per square foot. As a rule, the smaller the structure, the higher the square-foot cost. Package price for the 985-square-foot Cavendish, in the Gable series, is $27,100, or a bit over $27 per square foot. On the other hand, the 2,937-square-foot Wilmington B, also in the Gable series, at $52,700, works out to about $18 per square foot. For the precut logs and other components supplied by Real Log Homes for its most popular model, the three-bedroom Plymouth in the Gambrel series, you'd pay $26,600. Built and finished by a contractor, the house might end up costing in the neighborhood of $105,000. Do as much as you can yourself and you might bring it in for under $80,000.

Real Log Homes also offers log packages for Gambrel-style barns and for two- and three-car garages. If you're thinking *small*, there's always the 10 x 16-foot Cabot, package-priced at $5,900. The uses for this versatile little building are endless—hunting camp, playhouse, storage building, warming hut, or guest cabin. It's also ideal for those spots that are inaccessible to larger trucks. The precut logs and other components could be delivered by a 4 x 4 pickup or even a snowmobile.

Real Log Homes National Information Center
P.O. Box 202
Hartland, VT 05048
Phone: 1-800-REAL LOG
Market: North America, Japan, Europe
Product literature: plan book, $9; planning kit (plan book plus *Real Living* video), $20

Timberline Geodesics

Dome-shaped shelters, from simple bowed frames covered with animal skins to the igloo of the Eskimo, were among man's earliest constructions. But it wasn't until some thirty years ago that the mathematically precise geodesic dome was developed as a practical working concept.

R. Buckminster Fuller, who was the first to set down the math for constructing this geometric form—which combines the rigidity of the triangle with the strength of the sphere—envisioned the dome as the mass-produced home of the future. While still too radical for many, it has met with growing acceptance, and dome homes can be found today in nearly every section of the country.

The dome shape has a lot going for it. With a dome, it can take a third less structural material to provide the same amount of space as can be found in a home of conventional design. The dome can also take a third less energy to heat and cool. Since the entire weight of the dome, which is self-reinforcing, rests on its circular rim, there's no need for interior load-bearing walls. This allows for infinite design possibilities within. A dome can also be one of the easiest of all homes to erect.

There are two basic systems of dome building. One uses "space frames," which are large, prefabricated triangles that are simply bolted together to form the shell. The other is a system of steel hubs and wooden struts that are assembled as you go, with sheathing attached *after* the skeletal frame has been erected. For a number of reasons, not the least of which is that space frames are heavy and require a crew of four or five to position and bolt together, we prefer the more easily managed hubs and struts.

While all Timberline frames are essentially alike, the things that set one dome apart from another are the owner's choices in everything from roof covering and extensions to window placement and the entranceway. Explore all the possibilities. *(Timberline Geodesics)*

Domes can be built on everything from a concrete slab to a full basement. If you have a unique foundation need, as did the owner-builder of this home built into a slope, Timberline's engineers and architects can design one that will work for you. *(Timberline Geodesics)*

The best source today for dome kits employing hubs and struts is Timberline Geodesics. With Timberline, every hub flange and strut tip is color-coded, and even the largest of its domes can be easily assembled by two people with little or no previous construction experience. There's little tolerance for error with the frame of a geodesic dome, but then, with a system that's been continuously refined since 1969, it's almost impossible to go wrong. And unlike space frames, which can weigh 75 pounds and more, the heaviest component in a Timberline shell kit weighs no more than 25 pounds. The only tools needed to complete the shell of a Timberline dome are socket wrenches, hammers, ladders, scaffolding, and a nail gun. The latter two items are desirable but not essential.

Timberline offers two series of domes, a 3/8-sphere and a 5/8-sphere, and five diameters in each truncation: 24, 30, 35, 40, and 45 feet. Domes of less than 40-foot diameter are constructed atop riser walls to provide an 8-foot vertical rim and allow for the installation of doors. Where needed, riser-wall sections, preassembled and color-coded, come as part of the complete Timberline dome kit.

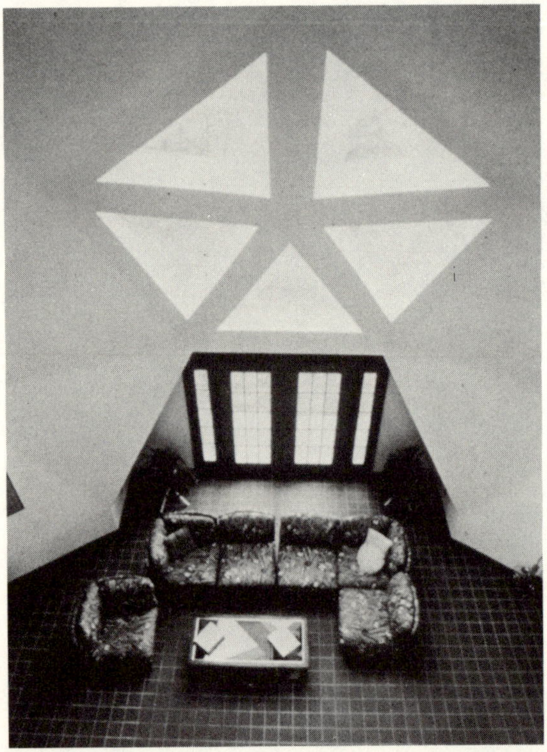

The clear-span structure of a dome shell opens up new ideas of how a home should be finished. Sample floor plans for a variety of lifestyles as well as photos showing how others have done it are included in the Timberline information package. *(Timberline Geodesics)*

One of Timberline's most innovative options is its patented Ledger Hanger. These can be coupled with hubs in the existing structural framework of any 5/8-sphere to hold as many ledgers as are needed around the perimeter to support the joists for a second floor or loft—eliminating any need for upright supports within the dome.

Each Timberline kit allows for two or more ground-level openings to accommodate doorways and other extensions. Up to five major openings are possible without in any way compromising the strength of the dome. Extension kits, which can be used to create a bay for a breakfast nook, utility room, or some other purpose, are obtainable as a Timberline option.

One of the many benefits of the natural symmetry of the dome is that it allows the placement of skylights and other windows virtually anywhere. If you are looking to achieve additional energy savings through passive solar energy, to take advantage of natural lighting, or to view the stars when in bed at night, you simply leave open those triangles where you plan to install skylights or any of the window variations offered as Timberline options.

Except with export packages, Timberline doesn't provide any insulating, roof-covering, or other finishing materials—all of which can be purchased locally for less than it would cost Timberline to bundle and deliver them.

There are three ways to buy a Timberline dome:

(1) The Complete Dome Kit, which includes beveled base plates and/or fully assembled riser walls; heavy-duty Timberline connectors; precision-cut 2 x 6-inch struts, color-coded and predrilled for connector bolts; studs and T-blocking for every panel; precision-cut plywood panels ready to be nailed up as

While Timberline's Complete Dome Kit includes only those materials needed to construct the basic shell, its comprehensive building guide takes you through everything from foundations and second-floor framing to deck, dormer, and roof plans. *(Timberline Geodesics)*

soon as you have erected the frame; all necessary nuts and bolts; three sets of your choice of Timberline floor plans, and the Assembly Manual.

Price: from $6,120 for the 24-foot 3/8-sphere to $16,545 for the 45-foot 5/8-sphere.

(2) The Strut Framing Kit, which includes all of the above with the exception of the precut triangular panels.

Price: $5,300 to $13,955.

(3) The Heavy-Duty Connector Kit, which includes the connectors—the heart of the Timberline dome assembly system—plus strut, stud, T-block, and base plate samples, and instructions on how to precision-cut and drill your own wooden components.

Price: $3,250 to $3,840.

How much does it cost to complete a Timberline dome? As a rule, a completed dome costs 10 to 15 percent less to build than a comparable box-type house. In addition, there's whatever you save by doing as much of the work as you can yourself—and this a system designed for the do-it-yourselfer. Timberline domes have been completed for as little as $40,000. Others have represented an investment of several hundred thousand dollars. In the end, it all comes down to how deep your pockets are and what it takes to make you happy.

Timberline's heavy-duty connector systems are available separately for those who prefer to cut their own lumber. Samples of struts, studs, T-blocking, and base plates, together with detailed cutting plans, are included with every connector kit. *(Timberline Geodesics)*

Timberline Geodesics
2015 Blake Street
Berkeley, CA 94704
Phone: 1-800-DOME-HOME
Market: worldwide
Product literature: catalog package, $12
Video: *Explore the Great Indoors,* $12

Shelter-Kit

The Unit One is the basic building block of a modular system of home construction introduced in 1970 by a New Hampshire craftsman named Andy Prokosch. Designed with the inexperienced owner-builder in mind, the idea was to produce a simple yet versatile shelter in kit form that could easily be assembled by two people in a few days, that could be later expanded by adding other modular components, and that could be kept low in price.

Under the name Shelter-Kit, the business expanded, much like many houses that begin as Unit Ones, and today offers not only the Unit One modular system, but modular Lofthouses, garages, and horse barns that are meeting with unexpected success as barnhouse conversions. All of these are offered in well-designed kits for construction by the owner-builder.

Shelter-Kit's basic Unit One module is a 12 x 12-foot weathertight shell of post-and-beam framing and with a sloping shed roof. There's

Designed for owner-builders with little or no construction experience, Shelter-Kit's modular Unit One (expanded here with a covered porch) can be assembled by two people in about 4 days. Kit includes everything needed for a weathertight shell. *(Shelter-Kit)*

Six 4 × 4 posts set on concrete piers support the floor and roof systems of the Unit One. Posts and beams, the key structural elements of the shelter, are predrilled at the factory and joined together with brackets and bolts at the building site. *(Shelter-Kit)*

Siding for the unit one is 1 × 8 pine tongue-and-groove clapboard. Studs of 2 × 4 spruce are used as nailing points for the siding and to support the framing for any windows. Every tool needed in the building of a Unit One comes with the well-designed kit. *(Shelter-Kit)*

also a 9 x 12-foot roofed platform, or porch module, that can be enclosed to make an additional room, and a 9 x 12-foot deck module. These can be combined, now or later, in any combination, to meet the owner's needs.

Foundation requirements with a Unit One are six poured-concrete footings or piers, one for each of the six posts that support the floor and roof systems and around which the walls are formed. Since the walls are non–load-bearing, it's possible to install a 12-foot sliding-glass door across the entire front of the unit and place other doors and windows in virtually any location. Post-and-beam construction also makes it possible to eliminate or shorten common walls when two or more Unit Ones are joined together.

The Unit One kit includes all the materials needed to construct a weathertight shell. All lumber is precut, predrilled where required, and labeled. All necessary tools and hardware are furnished. The package even includes two 6-foot stepladders and two carpenters' aprons!

The Unit One catalog illustrates a number of possible arrangements for everything from simple one-room cabins combining a Unit One with a porch and/or a deck module to three-bedroom plans calling for up to six Unit Ones and two decks. Price for a Unit One is $4,750. Price for the package calling for six Unit Ones and two deck modules is $25,610.

Since Unit Ones are often built in remote locations inacessible by anything larger than a 4 x 4, and sometimes not even that, Shelter-Kit packages the lumber and other components in bundles weighing no more than 100 pounds. All Shelter-Kit building packages are sold without insulation, partitioning, interior finishing materials, or fixtures.

If you are looking for an easy-to-assemble house of more traditional design, there's Shelter-Kit's expandable Lofthouse, which begins as a 16 x 16-foot or 20 x 24-foot peaked-roof module with a New Englandy look and a full loft. Preengineered extensions, lean-tos, porches, and decks can be added at any time. A Lofthouse 16 can be assembled by two people in about seven days and, with a total floor area of 512 square feet, is large enough to be used as a one-bedroom home. The 960-square-foot Lofthouse 20 takes about 14 days to erect and can accommodate two bedrooms.

2 -bedroom plan (629 square feet)

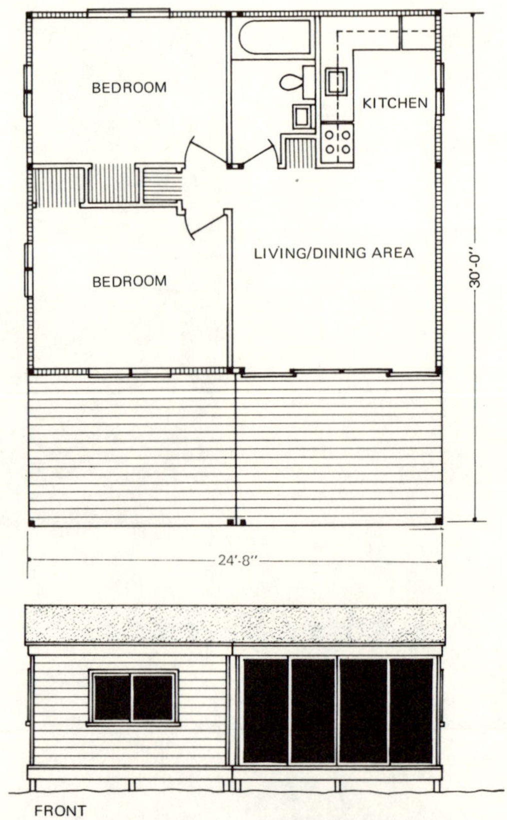

This two-bedroom plan, combining two Unit Ones, two enclosed porch modules, and two decks, costs $13,270 in preengineered kit form. *(Shelter-Kit)*

For about the same amount of money that it would take to build the two-bedroom plan shown on page 170, you could put up this more traditional two-bedroom Lofthouse, which combines the Lofthouse 16 with an 8 × 16-foot extension and a 10 × 24-foot deck. *(Shelter-Kit)*

With a few factory modifications, the basic kit for this sturdy yet stylish 24 × 24-foot horse barn has also become the kit for a surprising number of attractive two-bedroom residences. Others are finding it works very well as a garage, office, or workshop. *(Shelter-Kit)*

Either unit could make a good starter home, built with later expansion in mind.

As with the Unit One, framing for the Lofthouse is post-and-beam; the posts, though, are 6x6s rather than 4x4s. These houses can be built on concrete piers, concrete slabs, frost walls with crawlspace, or over full basements. Shell kits for Lofthouse 16 and 20 are priced at $8,970 and $14,150, respectively. The kits include everything needed to assemble a weathertight structure.

The newest innovation from Shelter-Kit is the Barn-House. It was never planned that way, but customers began asking for modifications to a kit that was originally designed to be a horse barn. They had found the barn design attractive and saw its potential for conversion into an inexpensive house.

You still can order the kit for a modular 24 x 24-foot post-and-beam horse barn (price, $8,000) that comes with an 8-foot rolling main door, and is roomy enough for two or three horses plus a hayloft, but the greater interest of late has been in the Barn-House conversion.

As a simple, low-cost structure for residential use, it's a wonder that the federal government and Habitat for Humanity haven't included the kits in their programs to house low-income families. Of sturdy, practical design, the post-and-beam Barn-Houses allow for a wide variety of floor plans and many options in door and window placement. The conversion includes the addition of a floor system (the barn provides for a dirt floor) and a change in the framing.

Available in two sizes, 24 x 24 and 24 x 32 feet, Barn-House kit prices are $11,060 and $13,745, with doors, windows, and stairs included. Both Barns and Barn-Houses are available in longer-length options.

Instructions for building Shelter-Kit Barns and Barn-Houses are written for the novice and include a primer covering everything from nail identification to such elementary techniques as toe-nailing and snapping a chalk line. If you read *and follow* the instructions that come with every Shelter-Kit structure, you're not likely to go wrong.

Shelter-Kit, Inc.
22 Mill Street, P.O. Box 1
Tilton, NH 03276
Phone: 603/934-4327
Market: nationwide
Product literature: catalog package, $6

Miles Homes

Miles Homes has helped more than thirty-five thousand families make their new-home dreams a reality. Catering to the do-it-yourself builder, the Miles owner-builder program lets you decide your level of involvement by building part or all of the home yourself, and subcontracting those portions you elect not to build. But what really sets Miles apart from others in the kit-house industry is its unique financial relationship with the owner-builder.

If you have the land on which to build, and "reasonably good" credit, Miles stands ready to help those willing to put in the time and labor with no-down-payment, below-market, interest-only construction financing. Monthly payments do not begin until sixty days after the first delivery of materials. Cash allowances may also be arranged for foundation, water-well, septic-system, and other preconstruction services the owner is not equipped to handle.

Not until the home is completed (Miles allows up to twenty-four months for this) and you have arranged for conventional, long-term bank financing, do you have to pay off your debt to Miles. By that time, depending upon your contribution to the building process, you may well have a home worth 50 to 80 percent more than the cost of its components.

With more than forty years of hands-on experience working with owner-builders, Miles's step-by-step instruction manuals, covering everything from the foundation to the final finishing, are possibly the best in the business. Miles also conducts regional owner-builder seminars and offers to its customers a video detailing the assembly process. Information on finding land, securing building per-

The popular Ponderosa model is one of Miles's country homes. Its features include three bedrooms, two-and-a-half baths, a Great Room, and a balcony overlooking the front entryway. *(Miles Homes)*

mits, and obtaining bids for subcontracted work is also available through Miles.

Once you have the foundation in place, Miles will deliver all materials needed to construct the shell of the house: floor system; framing lumber, precisely cut and coded as to location; sheathing; siding; roof system; doors and windows. Once you have the shell erected and weathertight, which represents about one-third of the work involved in the construction of the typical Miles home, Miles will make its delivery of insulation, drywall, and the electrical, plumbing, and heating systems packages. All nails and other fasteners are provided as needed.

To cut down on the time that it can take to complete the shells of its houses, Miles has recently turned to panelization on most of its standard home models. The company can now provide exterior and interior studwall sections (light enough to be handled by two people), as well as roof and floor trusses, in preassembled form.

Miles used to figure it took about 1½ hours of labor per square foot of floor area to complete one of its precut houses. But with panelization, significant savings can be achieved here, at least in the completion of the shell.

Using preassembled floor and roof trusses, panelized wall framing, and other preassemblies, one young Minnesota couple kept a log as they constructed the multilevel, four-bedroom, two-and-a-half-bath Foxwood, a 2,392-square-foot Miles design with a 528-square-foot attached garage. With the time they and others put in on the project, they found it took just over 600 hours total to complete the shingled, sided, and trimmed-out shell.

Spaciousness abounds in the South Hampton, a Miles home design featuring four bedrooms, two-and-a-half baths, dinette, and formal dining room. The first-floor mud room can also serve as a laundry room. *(Miles Homes)*

Miles's selection of homes designed to meet the needs of today's active families includes a variety of styles, from contemporary to colonial to Tudor and traditional, and more than seventy-five floor plans. But, in essence, your choice of home designs with Miles is virtually unlimited. Any of the Miles plans can be customized to meet your special needs. You can even bring in a plan from an outside source, and Miles will feed all the necessary details into its computers and deliver a complete precut package just as it would with any of its standard designs.

With the cost of each package based on customer specifications and other variables, the company is reluctant to quote prices. But kit prices on its basic designs would probably range from around $35,000 to $65,000. Including the cost of an average parcel of land, site prep, and other add-ons, Miles owner-builders have brought in four-bedroom homes like the Foxwood for under $90,000.

Be forewarned that not all who sit down with a Miles representative are approved for the program. Miles's success is directly related to its careful screening of applicants. You have to convince Miles managers that you have the ability and commitment to build your chosen house without a great deal of expensive professional help.

Miles Homes, Inc.
4700 Northern Lane, P.O. Box 9495
Minneapolis, MN 55440
Phone: 1-800-328-3380
Market: nationwide (except Hawaii and Alaska)

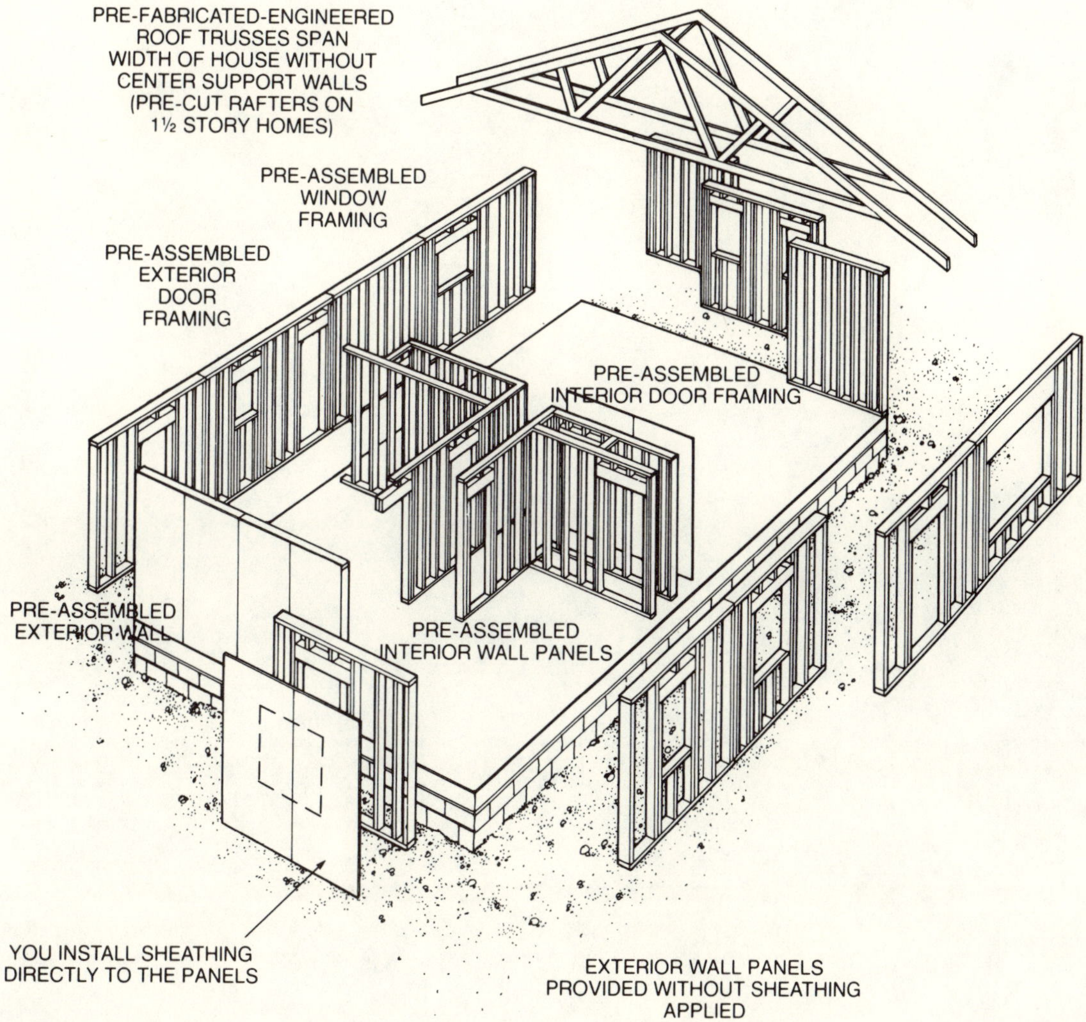

With panelization, Miles has made it less complicated for owner-builders doing their own framing and cut the time it used to take to erect the walls of one of its homes by up to 50 percent. *(Miles Homes)*

Transportation

AIRCRAFT

Liability problems and the enormous costs associated with gaining Federal Aviation Administration (FAA) certification for factory production have made the introduction of new, or even updated, single-engine piston aircraft prohibitively expensive for general aviation manufacturers. Nevertheless, the light-plane industry is thriving, and old-line manufacturers are being left in the dust by a host of companies turning out well-engineered kits for everything from ultralights and World War II fighter replicas to rotorcraft and sleek, high-performance, four-place cabin jobs.

Kit manufacturers avoid the formidable obstacles posed by the tort system and federal regulations by simply complying with the FAA's so-called 51-percent rule. Certification for production is not required and the kit manufacturer is more or less relieved of liability when the kit builder can be held responsible for more than half the work involved in the aircraft's construction. The government classifies such planes as "experimental/amateur-built" aircraft, restricts them to noncommercial use, and requires only a certificate attesting to their airworthiness.

The 51-percent rule is rather loosely applied. Prefabrication has reached the point where, with some kits, fuselages come from the manufacturer as two premolded halves that go together almost as easily as the snap-together components of a plastic scale-model airplane. A whole new cottage industry has also developed with proxy builders who have gone into the business of assembling aircraft for others who don't have the time or the shop skills to undertake or complete the job themselves.

It wasn't too long ago that kit builders could expect several visits from an FAA inspector during the construction of a home-built aircraft. Today, with limited manpower and the concurrent boom in homebuilts, there's only a single, often cursory, inspection made at the project's completion. Does this place the pilot of a homebuilt any more at risk than with a factory-built plane?

Many builders of kit planes today are experienced pilots to begin with and knowledgeable when it comes to flight-system mechanics. Being responsible for their own safety, they tend to take even greater care with the construction details of the aircraft they will be flying than an assembly-line worker might. Records kept by the National Transportation Safety Board bear this out. The vast majority of light-plane accidents can be attributed to lack of pilot experience, and the strongest recommendation we can make is to take pilot training with a license instructor in the particular type of aircraft you plan to be flying. Even with unlicensed ultralights, don't even think about going aloft in the craft you build without prior training.

Ultralights can be fun and are one of the least-expensive ways to get airborne, but the hot kit planes today are the sleek and sexy composites, with bodies assembled from major, prefabricated fuselage and wing sections. Composites are a space-age lamination of epoxy, layers of (usually) woven fiberglass cloth, and a rigid foam core. Kevlar and carbon fiber are also being used. Weight for weight, composite airframes are much stronger than the more common types of small-plane construction, such as aluminum and wood, and steel tube and fabric.

If you are looking for high performance or a snazzy four-passenger aircraft, a composite would almost have to be your first choice. There are both two- and four-seat composites that can blow by almost any single-engine aircraft coming from Piper, Beech, or Cessna today. And these planes can be assembled from mostly preformed parts for less than half the cost of most factory-builts.

Composites are more costly than other kits, but they can also cut assembly time significantly. It *does* take time to construct an airplane from a kit; as much as 2,000 hours with some kits. That's 40 hours a week for 50 weeks, or 20 hours a week for two years! Not all kit planes take nearly that long, but then some assemblers are in it as much for the joy of building as for the joy of flying. For those seeking more immediate gratification, there *are* kit planes that can be assembled in under 100 hours. It's a well-known fact, though, that a lot of kits never do get completed. While the introduction of composites has brought about a substantial increase in the rate of completions, the key here is to settle for an aircraft that does not overmatch your construction and flying skills. You'll also need a spacious, comfortable work area.

Kit prices, which range from a few thousand dollars to $50,000 and more, can be deceptive. Some kits include only the airframe. Others are priced to include an engine and propeller as well. Still others add basic instrumentation. With a kit costing $20,000 or more for the airframe alone, you might easily have to spend another $20,000 to make the plane airworthy. But then, as will be seen, there also are aircraft that can be completed for less than half the cost of the average new car.

If you *are* serious about building and flying your very own plane, a good first step would be to join the 129,000-member Experimental Aircraft Association, which has local chapters nearly everywhere. Shared experience with others of the same mind can be invaluable. The address of the EAA is 3000 Poberezny Road, Oshkosh, WI 54903-3086. The goal of virtually every homebuilder we've known, even though the project may have been years in the making, was not only to complete his or her aircraft but to fly it to the mecca for homebuilts—the annual midsummer EAA Oshkosh Fly-In Convention, which fields upward of 15,000 transient aircraft and attracts more than 800,000 attendees.

We have been aviation enthusiasts since our early youth, and that was when pilots were known as aviators. We've also built our share of wing ribs and doped enough fabric to cover a tennis court, but we are in no position today to tell anyone what type of aircraft they should build—not with some 175 kit-plane companies offering nearly four hundred models of often wildly different design, price, and degree of difficulty of construction. We can only point the reader to a selection of companies that have been unusually successful with their particular type of aircraft, a good indication that both the product and the price have been right.

Stoddard-Hamilton Aircraft

In 1979, Stoddard-Hamilton took sports aviation into a new era with the introduction of

Sleek Glasair two-seaters, in a choice of fixed or retractable undergear configurations, are setting the standards by which all other high-performance composite homebuilts are judged. *(Stoddard-Hamilton Aircraft, Inc.)*

the Glasair, the first completely premolded composite kit aircraft. Today, with over twelve hundred kits sold and a choice of five models designed to suit the performance desires and budgets of a wide range of pilots, Glasair two-seaters continue to set the standards by which all other composite homebuilts are judged.

The Glasair was first introduced as a conventional fixed-gear taildragger (TD) and later additionally offered with a steerable nose wheel for both fixed trike (FT) and retractable gear (RG) undercarriages. Constant refinement of an already outstanding product led to a second generation of Glasairs, the Glasair II-S (for stretched) class, in all three undergear configurations. The II-S series was followed by the Glasair III and the extra-high-performance Glasair III Turbo, which are available with retractable gear only.

All Glasairs are high-performance aircraft, designed primarily to be fast, efficient cross-country machines, and secondarily to provide above-average aerobatic capability. Glasair II-S models, equipped with a choice of 4-cylinder Lycoming aircraft engines, from 150 to 200 horsepower, fill the 190- to 250-mph performance niche. The Glasair III is designed to use a 6-cylinder Lycoming, from 250 to 300 hp. It can cruise at 282 mph and hit a high of 295. For those with a passion for absolute speed and high-altitude flight, the turbocharged Glasair III cruises at 313 mph and can push 330 at 24,000 feet.

From the sporty fixed-gear Glasairs to the sleek retractables, the Glasairs' respectable short-field capability, excellent cruise speeds, and long range will open up nearly limitless airfields and destinations for the proud owner of any one of these homebuilts to discover. Seating is side by side in the two-seaters, which are roomy enough to accommodate a pair of 6-foot-6, 200-pounders. Glasairs also boast the largest baggage compartment in the industry for two-place aircraft. Wingspan of all five models is 23.3 feet. Lengths range from 20.2 feet for the II-S craft to 21.3 for the Glasair IIIs.

Fabricated from preformed shells made in female molds by Stoddard-Hamilton, Glasair airframe components arrive with a beautifully smooth, primered finish ready for final paint-

Glasair kits are some of the most comprehensive in the industry. Preformed composite airframe components arrive with a beautifully smooth, primered finish ready for final painting after assembly. *(Stoddard-Hamilton Aircraft, Inc.)*

ing after assembly. Assembly is very much like putting together a giant plastic model-airplane. The fuselage, for example, comes in two large half-shells and three belly sections.

Estimated building times for Glasairs range from 1,200 hours for Glasair II-S fixed-gear designs to 1,800 hours for the Glasair IIIs. Planes of this caliber built with aluminum tube and fabric or wood would take at least twice as long to construct. Glasair kits include all parts necessary to complete the airframe and can be purchased as a single package or as four component kits. The System 4 purchase plan spreads the payments and reduces the kit builder's storage and workspace re-

Offering blistering cross-country speed and above-average aerobatic capability, Glasair IIIs, with or without the war paint, look and handle a lot like World War II fighter aircraft. *(Stoddard-Hamilton Aircraft, Inc.)*

quirements. Not included in the kits are such off-the-shelf items as the engine, propeller, paint, instruments, avionics, and related hardware and controls.

Kit prices range from $16,900 for the II-S TD to $33,500 for the Glasair III. Typically, a Glasair II-S FT, with a kit price of $17,500, a used engine ($3,500), a basic VFR panel, and a mixed bag of new and used avionics and instruments ($4,000), could be completed for under $28,500. At the other extreme, a Glasair III, with a new factory engine, new IFR equipment, and the interior and paint job professionally done, would total approximately $85,000. With a mid-time engine, a basic non-IFR panel, and do-it-yourself painting and interior finishing, this could easily be reduced by some $25,000.

Glasair provides a strong support system for its kit builders. Technical assistance is never more than a phone call away, both during and after construction. The company also publishes a thick quarterly newsletter filled with builder hints, safety tips, letters from Glasair owners, and other items of interest.

Each of the Glasair models has different flight characteristics, requiring varying degrees of pilot experience and ability. So that kit buyers don't outmatch themselves, Glasair offers familiarization rides with a factory rep in demonstration models prior to a purchase. A similar program is offered to builders who are within three months of completing their own Glasair, to give them experience flying a factory Glasair prior to that first test flight in their own aircraft.

Should the builder of a Glasair ever have thoughts of selling the completed aircraft, there should be some comfort in knowing that Glasairs probably command the highest resale value among two-place homebuilts. The documented average resale value of the Glasair II-S RG, with 160 or 180 hp, is $75,000. The average resale value of a completed Glasair III is $136,000.

Stoddard-Hamilton Aircraft, Inc.
18701 58th Avenue N.E.
Arlington, WA 98223
Phone: 206/435-8533
Product literature: 60-page Glasair information pack, $20; 48-minute video, $25; or both info pack and video, $40

Denney Aerocraft

With more than fifteen hundred kits already out there, Denney Aerocraft's Kitfox, which has been licensed to fly in over twenty-four foreign countries, has become the most popular homebuilt aircraft in the world. The Kitfox first flew in 1984 and has gone through three model changes while still retaining its classic 1940s slow-riding bush plane image. As a rugged two-place taildragger with short takeoff and landing (STOL) capabilities, and flying speeds of from 35 to 115 mph, its attractions are a far cry from those offered by the high-performance composites.

A blend of nostalgia and new technology, the Kitfox was designed for fun flying and the first-time builder. It uses auto gas and is a near-foolproof plane both to build and to fly. Construction is traditional tube and fabric, with all the hard work—from the prewelded fuselage skeleton to the fiberglass cowl parts—done at the factory. The aircraft can be assembled in a one-car garage and takes about 600 man-hours to complete, with no special skills or tools required.

The complete package for the Kitfox IV ("the best just got better"), including a 65-hp

Designed for flying fun and the first-time builder, Denney's rugged little Kitfox, with its classic 1940s bush plane image, has become the most popular homebuilt aircraft in the world. *(Denney Aerocraft Company)*

dual-ignition Rotax 582LC two-stroke engine and a few essential options, such as an altimeter, a compass, and a tachometer, is $16,243. You could get by for that, but then there are some other options you may find hard to resist: a bigger engine for greater cruising speed, skis and a heater for wintertime fun, amphibious floats, wing tanks for extended flying time, full Lexan doors for greater visibility, and classic-design wheel pants. Choose the $4,872 engine upgrade, the $6,995 amphibious floats, a telescoping oar, and some avionics, and you're looking at a $30,000 package.

It's this wide variety of options, however—and there are four pages of them—that allows the builder to tailor the Kitfox to his or her particular needs and wants. Although most builders order the complete kit for shipment at one time (it comes in one *big* box), others

The complete kit package for the Kitfox IV, including a 65-hp Rotax two-stroke engine and a few essential options, such as an altimeter, a compass, and a tachometer, costs under $16,500. *(Denney Aerocraft Company)*

A wide range of options is offered with the Kitfox IV, from factory-made floats and skis to wheelpants, extra fuel tanks, and variations in upholstery colors and instrument panels. *(Denney Aerocraft Company)*

opt to receive it in five stages: fuselage kit, wing kit, landing-gear kit, engine kit, and covering-and-finishing kit. The payments can be spread, but the total costs run a little higher because of additional packing and shipping charges.

One of the most time-consuming tasks in building the Kitfox is preparing the prewelded fuselage and other metal parts—from struts to door frames—for painting. Sandblasting can speed the work, but Denney Aerocraft now offers powder coating, a $700 solution to the sometimes difficult task of preparing and painting the metal work properly. Denney's powder-coating option includes filing and smoothing all surfaces where needed before the color coating is applied. Because the powder coating, offered in a choice of nine stock colors, is electrostatically bonded to the metal parts and then fused in an oven, a coating is produced that is virtually impossible to match with spray or brush applications.

Most Kitfox builders will not have had any experience with fabric covering and may be apprehensive about this stage of the project. However, covering the airframe is not difficult, and the construction manual takes the builder through the process step by step. There is no rib stitching or other sewing. All fabric (Dacron) is attached with special glue and heat-shrunk (made taut) with an iron. The entire Kitfox can be covered in three to four weeks, working part time.

For the finish, Denney recommends the minimum-weight, three-coat Stits Poly-Fiber Coating process. Using this process, the airplane will weigh 20 to 40 pounds less than it would if a nitrate/butyrate doping process is used. The Stits system consists of one coat of Poly-Brush sealer (included in the kit) and two color coats of Poly-Tone with an ultraviolet (UV) blocking additive. It takes approximately four gallons of Poly-Tone to cover the entire aircraft. This can be divided into several colors if a mixed color scheme is planned.

Its 130 square feet of wing area and full-span flaperons enable the nimble Kitfox to be airborne in 75 feet, climb at 1,300 feet per minute, land at less than 30 mph for a 100-foot rollout. *(Denney Aerocraft Company)*

The Stits paint process helps to keep the Kitfox at its targeted empty weight of 440 pounds—giving the aircraft a useful load weight of up to 610 pounds.

The nimbleness of the Kitfox is seen in its ability to be airborne in 75 feet, which is less than five times its own length of 17.7 feet. The lift provided by its 32-foot wingspan, and 130 square feet of wing area, enables it to climb at better than 1,300 feet per minute. When landing, full-span flaperons can slow the plane to less than 30 mph for a 100-foot rollout. Another of the Kitfox's attractions is that its wings can be folded back in less than five minutes for easy towing. Raise the tail, attach the aircraft by a towbar to the car, and it can be trundled home on its mains for garaging, saving airport hangar fees.

Denney Aerocraft Company
Nampa Airport
100 North Kings Road
Nampa, ID 83687
Phone: 208/466-1711
Product literature: Kitfox information pack, $10; video, $15

RotorWay International

The two-place Exec 90 helicopter is the product of nearly three decades of engineering, flight testing, and refinements. Introduced under new ownership (RotorWay International) at Oshkosh in 1990, it's a significantly advanced version of the original Exec helicopter, which, through the 1980s, had been the most popular homebuilt helicopter ever flown. The Exec 90 bids fair to succeed to that title.

Pilot and passenger will find plenty of leg, elbow, and head room in the enlarged "cabin" of the Exec 90, which has a useful load weight of 500 pounds, a 22-foot-long fuselage, and a 25-foot rotorspan. Running on 92-octane auto gas, the sport copter cruises at 90 to 100 mph, and has a top of 125. Its 17-gallon fuel tank provides for about two hours of flying or a 180-mile range.

Everything needed to assemble the Exec 90 comes with the kit, including a rugged, RI 162-cubic-inch, liquid-cooled, dual-ignition power plant. This durable and thoroughly tested 4-cylinder engine was designed and de-

Running on 92-octane auto gas, the two-place Exec 90 gives new meaning to sport flying—with access to areas and an infinite number of landing zones denied to pilots of fixed-wing aircraft. *(RotorWay International)*

veloped specifically for helicopter use by RotorWay, which has its own foundry and manufactures the engines in its own plant. Unlike almost all other "complete" aircraft kits, the Exec 90 kit *does* include virtually everything you need, right down to seat belts and basic flight instruments. About all the kit builder has to provide, other than labor, is paint and gas.

Every nut, bolt, washer, and cowl screw that goes into the helicopter airframe is there, clearly labeled and secured with shrink wrap. Many of the critical subsystems are already

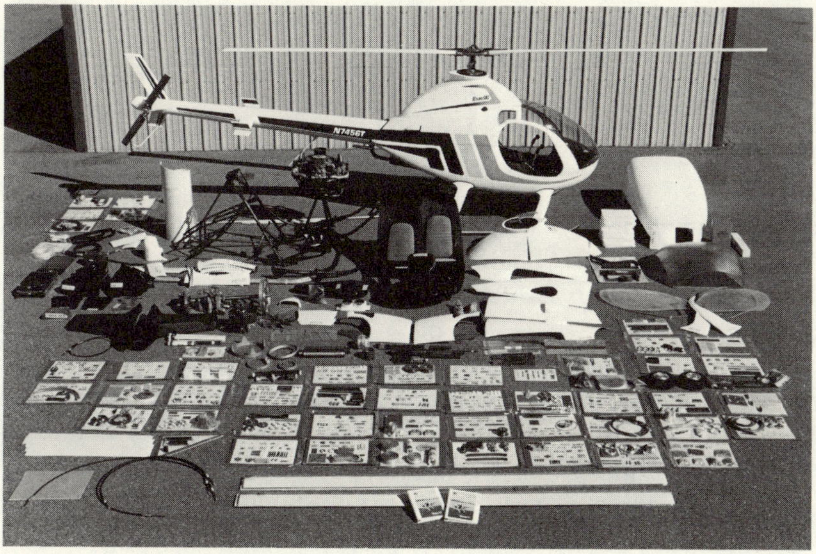

The Exec 90 kit includes everything needed to get to the flightline. But even before you get there, the Exec 90 program will teach you how to fly your copter and how to maintain it. *(RotorWay International)*

fabricated, ready to bolt on. The rotor system, for example, comes completely assembled. All of the structural welding and machining is done at the factory. Fuselage skins, fairings, cowling, and the acrylic windscreen are all preformed, guaranteeing an exact fit with a smooth-as-glass finish. Even the rotor blades require little more than finishing touches.

A "see-do" manual guides the kit builder through every stage and assures that even first-time aircraft builders can progress through the program without getting lost. For most people, the program for getting from the uncrating and parts inventory, through the step-by-step construction to the flightline, takes about 500 hours, or from six months to a year of part-time work. That's with the standard kit, which carries a kit price of $42,500. For a little more money, there's also a new Quick Kit that can get the builder of an Exec 90 to the flightline about 250 hours faster.

As an essential part of the construction process, RotorWay conducts a two-week program at its Arizona base to train builders of Exec 90s how to fly and maintain their aircraft. RotorWay has a flight examiner on staff, and the program leads to a private helicopter license or a helicopter rating add-on for those already ticketed to fly fixed-wing aircraft. One of the benefits connected with building an Exec 90 is that it qualifies the builder for an Aircraft Repairman's Certificate. With the certificate and the training program, builders of the Exec 90 are then authorized to do all of their own maintenance. This could, in time, save thousands of dollars.

Depending on where they live, Exec 90 owners might be able to fly from a backyard heliport. As an alternative to paying airport hangar fees, the Exec 90 can also be transported on a simple, snowmobile trailer and stored in a one-car garage. It takes about an hour for one person to attach or remove the rotor blades, and a set of wheels slipped onto the skids of the Exec 90 makes ground-handling easy.

RotorWay International
300 South 25th Avenue
Phoenix, AZ 85009
Phone: 602/278-8899
Product literature: Exec 90 info pack, $15

Loehle Aviation

The P-51 Mustang was one of the best-ever piston-engine fighter planes and was credited with the destruction of nearly ten thousand enemy aircraft during World War II. Noted for its 400-plus-mph speed, it packed six 50-caliber machine guns and was the preferred war bird of most young American fighter pilots of the time.

It's still an immensely popular plane with pilots both young and old, and at least five manufacturers offer P-51 replica kits. A couple of those kits are for high-performance aircraft costing $50,000 or more and with an estimated construction time of from one to three years. That calls for an awful lot of nostalgia as well as dedication on the kit builder's part. Loehle's "5151" Mustang, our choice in a fighter replica, costs under $10,000 and can be built in 400 to 500 hours.

The Loehle three-quarter-scale Mustang doesn't pack any guns and, with its 50-hp Rotax 503 engine, has less than one-fourth the speed of the original. But it does have the right look, right down to the P-51's characteristic belly scoop. There's also a retractable-gear option ($1,554), which most buyers find hard to pass up.

The basic kit for the Loehle "5151," a three-quarter-scale replica of the P-51 Mustang, one of the most celebrated of World War II fighter aircraft, costs under $10,000, engine included. *(Loehle Aviation)*

Construction of the "5151" is essentially wood and fabric, with most wood and metal parts prefabricated and numbered for easy assembly. The retractable landing gear is a popular option. *(Loehle Aviation)*

Construction of the "5151" (that was also the price of the kit when it was first introduced some 125 kit orders ago) is essentially wood and fabric. Most wood and metal parts are prefabricated and numbered for easy assembly. The fuselage is the well-proven deHavilland-type box, with laminated spars and geodetic bracing. All wing ribs are precut and profiled to form the airfoil taper.

As with the Kitfox (see pages 181–84), the airframe of the "5151" is covered with Dacron, which is glued on and then heat-shrunk to drum tightness with a common household iron. The Stits lightweight sealing-and-coloring process is again recommended. For a paint scheme, most Mustang builders apply some version of WWII war-bird plummage.

The basic kit for the "5151" Mustang with the 50-hp Rotax costs $9,995. Dope, paint, and instruments are not included. An optional engine upgrade, to a 65-hp Rotax 582LC, would bring the kit price to $12,140. For those on a limited budget and anxious to get started, Loehle makes the "5151" available as eleven sub-assembly kits. The option allows you to pay as you build.

Empty weight of the aircraft with the Rotax 503 is 513 pounds, gross weight is 885. The wingspan is 27.4 feet, length is 22.8 feet. The "5151" has short-field capabilities, with a takeoff roll of 150 feet. It cruises at 75 to 85 mph, stalls at a very low 30, and has a top of 100. For ground transport and storage, the wings and tail of the plane can easily be removed.

Loehle Aviation, Inc.
"The Aviation Valley"
Shipmans Creek Road
Wartrace, TN 37183
Phone: 615/857-3419
Product literature: info pack, $10; video $25

TEAM, Inc.

The ultralight movement got its start in the late 1970s when an inspired birdman added a go-kart engine and propeller to his hang glider. Another birdman hung a tricycle from his glider and mounted a chain-saw engine on it. In no time at all, the sky was buzzing with unlicensed engine-driven flying wings, and the FAA stepped in to give definition to this new class of flying machines.

There are restrictions on where, when, and how fast these aircraft can fly today, but the real defining line comes with the imposed empty-weight limit of 254 pounds, which the FAA set to separate ultralights, for which no pilot license is needed, from experimental/amateur-built aircraft, for which a private pilot license *is* needed.

To come in under the weight limit, a lot of ultralights continue to resemble flying tricycles, and, just as in the early days of aviation, flying one is pretty much by the seat of one's pants. That's where TEAM's very successful mini-MAX and hi-MAX differ. While these diminutive single-seaters qualify (with empty weights of 237 and 250 pounds, respectively) as legal ultralights, they look and handle a lot like conventional aircraft.

Designed from the outset to be powered by a 27-hp Rotax engine, the prototype of the mid-wing mini-MAX first flew in 1985. Offered as a kit, and constructed much like a wood-and-tissue model aiplane, it immediately became popular with light-plane enthusiasts throughout the world. TEAM doesn't stock or supply any engines for its all-wood, fabric-covered aircraft, but Rotax 277 and 377 engines, as well as VW conversions, have been popular choices. At a kit price of $2,695, and requiring about 300 hours of construc-

(*above*) At 237 pounds, the mini-MAX qualifies as a legal ultralight, but bears little of the usual resemblance to a flying tricycle. Instead, it looks and handles a lot like a conventional aircraft. *(TEAM, Inc.)*

(*below*) Constructed much like a wood-and-tissue model airplane, the all-wood, fabric-covered mini-MAX takes about 300 hours to assemble. It can be brought in, with engine, for under $4,500. *(TEAM, Inc.)*

The hi-MAX is a high-wing, enclosed-cockpit version of the mid-wing mini-MAX. Both models have a useful load weight of 260 pounds, a top speed of 60 mph, and land (with flaps) at 22 mph. *(TEAM, Inc.)*

tion time, a mini-MAX can be brought in, with engine, for under $4,500. The kit for the hi-MAX, which is a high-wing, enclosed version of the mini-MAX, costs $500 more and takes about 50 hours longer to construct.

Wingspread of both models is 25 feet, and length is 15.5 feet. The planes have a useful load weight of 260 pounds, cruise at 55 mph, with a top speed of 60, and land (with flaps) at 22 mph. Takeoff is quick, requiring only about six seconds and a ground roll (on grass) of less than 150 feet.

TEAM, Inc.
Box 338C - Route 1
State Road 53 & Ivy Bluff Road
Bradyville, TN 37026
Phone: 615/765-5397
Product literature: information pack, $5; video, $19.95

BOATS

Just about anyone with average do-it-yourself skills can build a boat from a precut kit, or even a partial kit that might include little more than the framing members that give shape to the hull. Since a boat largely handcrafted from a kit can represent a savings of from 40 to 50 percent of what a comparable factory-built model might cost, the newer, improved kits are becoming an increasingly popular alternative with would-be boat owners. It's also the only way to know a boat inside and out and be assured of quality every step of the way.

Whether you've been hankering to cruise, sail, or troll for the big ones, or just to jink about on a pond in a windblown pram, you'll find there are kits for every preference and budget, representing everything from kayaks and canoes to sailboats and cabin cruisers. While the kit boats selected for coverage here are essentially wooden boats, the disadvantages often associated with wooden hulls have been eliminated, largely through the use of premium epoxy resins and fiberglass hull coverings. The direction being taken today is toward a low-maintenance "composite" hull—a sandwich of epoxy, wood, and fiberglass, with wood as the meaty core.

The two leading mail-order sources of do-it-yourself watercraft kits, Clark Craft and Glen-L Marine, have both introduced epoxy encapsulation systems that, applied to the bare wood of a boat, seal the wood against marine borers, rot, and the other deleterious effects of moisture, such as warping, delaminating, shrinkage, and swelling. The epoxy also strengthens the wood significantly and, when thickened and used as a glue or putty in the assembly of a hull, can form a bond that is actually stronger than the components joined.

It's this latter characteristic that has led to a proliferation of "stitch-and-glue" boat kits. Here, the hull is literally folded into shape and held together with copper stitches to give the epoxied seams time to cure. The entire hull may also be epoxied, inside and out, and/or given a fiberglass covering for added strength and a smooth, professional finish.

We know that many of you, beginners especially, are hesitant about using fiberglass and resins. But this is a far cry from building a mold and laying up a fiberglass hull. In the following pages, we'll take you through several of these processes so that you'll know exactly what's involved. The work could be a lot easier than you may have been led to believe.

WoodenBoat

Our choice in an all-around dinghy, The Nutshell Pram, comes with excellent credentials. The folks behind it publish *WoodenBoat*, the only magazine in the world devoted exclusively to boats built of wood, and conduct the WoodenBoat School, which, between mid-June and October of each year, offers more than thirty-five boatbuilding courses.

Since its introduction in 1983, hundreds of Nutshell Prams have been built, and they've become a common sight in U.S. coastal and inland waters. For this is one pram that rows as well as it tows, sculls as well as it sails. It's a fine, forgiving little craft, and a real joy for both beginner and experienced sailors or oarsmen.

The Nutshell Pram is an all-around dinghy that sails as well as it sculls. With its lapstrake plywood construction, it's a simple boat to build, in either rowing or sailing versions. *(WoodenBoat Publications, Inc.)*

With its lapstrake plywood construction, the Nutshell Pram is a simple boat to build. It can be constructed in either of two sizes: 7 feet 7 inches or 9 feet 6 inches. The larger craft is not only 2 feet longer but 4 inches beamier than the original Nutshell Pram. It carries a bigger load (more gear or more people or both), has more sail power, and is more stable.

Kits, available in either rowing or sailing versions, include all wood parts rough-cut to size, fastenings, hardware, oarlocks and sockets, plans, and step-by-step illustrated instructions. Sailing models include parts for the spars, rudder and fittings, tiller, daggerboard, and case.

Prices start at $700 for the smaller rowing model, $850 for the sailing model. Add another $195 for an optional Dacron sail, in either white or tanbark. Kit prices for the larger model are $950 and $1,150, respectively, plus $295 for the sail.

Before placing an order for a Nutshell Pram, you can preview the construction process. WoodenBoat offers a 90-minute video, *How to Build the Nutshell Pram*, which takes you from the start to in-the-water demonstrations of rowing, sculling, and sailing techniques. The video can be purchased for $59.95 or rented for thirty days for $11.95.

An alternative way to learn the basic plywood construction techniques would be to warm up for the task ahead by building WoodenBoat's scale model of the 7-foot-7-inch Nutshell Pram. At a scale of 1½ inches to 1 foot, it's true to the features of the larger

Building a scale-model (kit price, $39.95) of the Nutshell Pram is a great way to learn the basic plywood construction techniques if you're thinking about building a full-size Nutshell Pram. *(WoodenBoat Publications, Inc.)*

craft. The Nutshell Pram scale-model kit costs $39.95.

For scratch builders, plans, including full-size patterns, are available for either of the full-size Nutshell Prams at $75.

The WoodenBoat Store
P.O. Box 78
Brooklin, ME 04616
Phone: 1-800-225-5205

Glen-L Marine Designs

Catering to the amateur boatbuilder, with plans, full-size patterns, and a wide variety of kits, Glen-L Marine Designs has been a leading name in mail-order do-it-yourself watercraft for more than thirty-five years. The Glen-L catalog offers some 189 designs, covering everything from a 7-foot Stitch-N-Glue dinghy and an 8-foot hydroplane to ski boats, bass boats, sloops, catamarans, and trailerable 27-foot cruisers. If Glen-L doesn't have it, you probably don't need it.

Rather than offering full-component kits, Glen-L holds down the cost of its kits by requiring builders of its boats to purchase locally the standard lumberyard longitudinals and plywood needed to plank the hull and deck. Glen-L then reduces the possibility of error on the boatbuilder's part by offering precut frame kits for many of its designs. Machined using master templates and assembled in factory jigs, the framing members, when delivered to the kit builder and set up as detailed in the plans, form the basic contoured "backbone" of the boat. It's around this that the longitudinals are sprung and the planking is applied. Glen-L also provides fastening kits, in either bronze or hot-dipped galvanized steel; fiberglass covering kits, in either polyester or epoxy resin; and, for sailboats, hardware, rigging, sail, and aluminum spar kits.

Typically, for the Glen-L 10, a 10-foot cat-rig sailboat with daggerboard, the frame kit ($210) includes all frames, preassembled; the fully formed transom; stem and mast-support assembly; bowpiece; deck beam; rudder; daggerboard; daggerboard trunk; complete plans with instructions; bill of materials, and a fastening schedule. A fastening kit, in bronze, would add about $44, and a hardware kit, including cleats, blocks, and a kick-up rudder fitting, would add another $100. Figure $90 for the fiberglass (epoxy resin) hull-covering kit; $125 for the aluminum spar kit, including the mast and fittings; $140 for the rigging kit, and $200 for a low-maintenance Dacron sail and matching sail bag. Including the sheet plywood purchased locally, you've got yourself a full-fledged sailboat with an investment of less than $900.

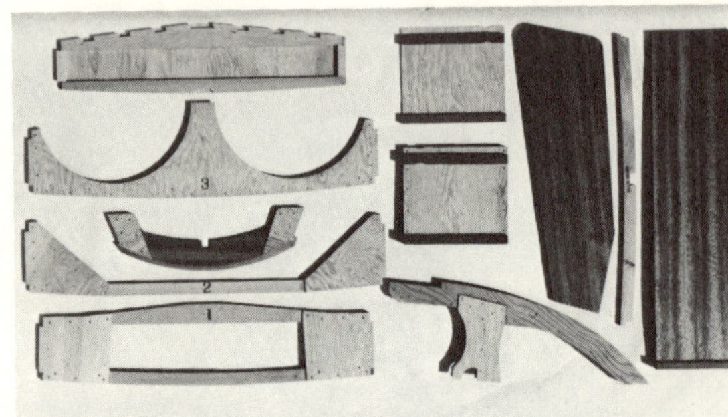

The factory-prepared frame kit for the Glen-L 10 eliminates potential errors and saves assembly time, inspiring confidence early in the construction when it is most needed by the beginner. *(Glen-L Marine Designs)*

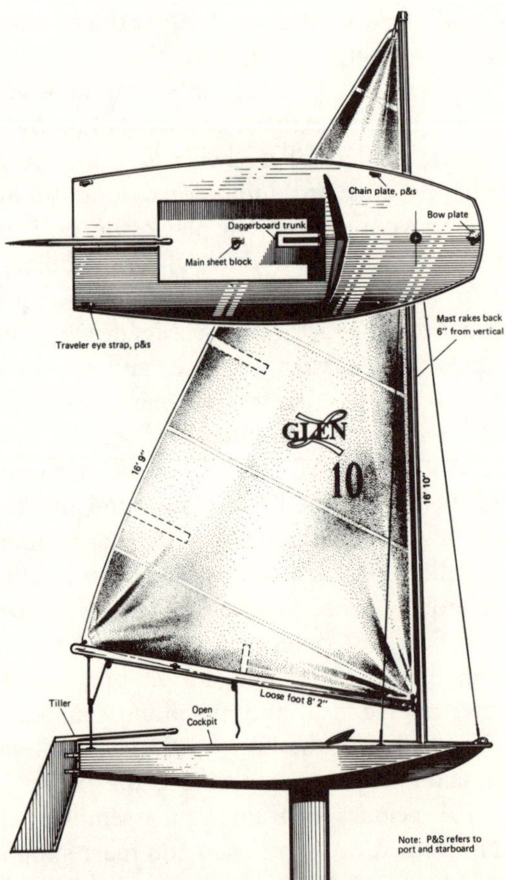

A full-fledged cat-rig sailboat with an investment of less than $900, that's the Glen-L 10. All aspects of the design have been kept simple, making this an ideal sailboat for the beginner. *(Glen-L Marine Designs)*

The Glen-L 10 is very lively under sail, points well, and is highly maneuverable. The lightweight hull fits easily into most vans or pickup beds, or can be cartopped. All aspects of the design, from rigging to sailing, have been kept simple, making this a practical sailboat for the beginner.

Glen-L's most complete kit is reserved for its 9½-foot Picklefork three-point hydroplane. While not meant as a competition hydro, it is very much like the racing pickleforks found on the competition circuits, only in more compact form. A frame kit ($227) is available, but Glen-L also offers a "complete" parts kit ($365), which includes all precut spruce, mahogany, and plywood parts, with all components numbered for identification and ready for assembly. The kit is an excellent choice for the beginner who doesn't want a first boatbuilding project to be anything other than quick and easy. Allow another $25 for fastenings in hot-dipped galvanized steel, or $65 in bronze, and $75 for a fiberglass hull-covering kit in polyester resin, or $155 in epoxy resin. Power supplied by a 7.5-hp outboard motor would add $1,000 or so to the final cost of the Picklefork.

Named for its twin-point prow, the Picklefork three-point hydroplane represents Glen-L's most complete kit, with all parts preshaped, numbered for identification, and ready for assembly. *(Glen-L Marine Designs)*

For a leap ahead in simplified boat construction, there's Glen-L's 12- and 14-foot Power Skiffs, which are assembled using the Glen-L Fold-And-Stitch-Then-Glue (FAST-G) method of construction. The special Stitch-N-Glue Power Skiff frame kit ($147) includes transom, motorboard, filler, ledge, stern knees, transom knee, breasthook, center form, center seat, and a full set of plans and patterns for building the model chosen. All it takes in the way of plywood to build the 12-footer is three 4 x 8-foot sheets of ¼-inch plywood and a half-sheet of ¾-inch stock. Transfer the patterns to the plywood, cut out the hull components, drill holes for stitching every 4 inches along the edges to be joined, and lay the parts out on the ground like a Mercator (map) projection. Fold the parts together progressively and stitch and glue the seams—and you have the hull.

The stitching is done with short lengths of copper wire passed through pairs of parallel holes along the seam edges and twisted on the hull exterior to hold the panels together. The glue is an epoxy putty formed in a cove-shaped fillet along the seams, usually on the inside of the hull. The stitched-and-glued seam is then covered with strips of resin-coated fiberglass tape on both the inner and outer surfaces of the hull, but not before the copper twists are snipped and filed flush with the hull exterior.

The Stitch-N-Glue kit ($160) for the Power Skiff contains all materials required for stitching and fastening the hull components together and to epoxy-encapsulate the interior. The kit includes epoxy resin, copper stitching wire, fiberglass tape, rollers, brush, and all needed screws and carriage bolts. The fiberglass kit for covering the exterior of the skiff costs another $150.

The unique hull design of the Power Skiff features a hard chine-arc bottom aft and an easy-riding convex-vee keel entrance forward for agility, stability, and speed. Built-in flota-

The Stitch-N-Glue method of boatbuilding consists of cutting the planking panels to shape, assembling them on the ground, and then stitching and gluing the panels together at the seams. *(Glen-L Marine Designs)*

The boat, in this case the Glen-L 12-foot Power Skiff, folds together and is wire-stitched at seams which later will be filleted with thickened epoxy resin and taped with fiberglass. *(Glen-L Marine Designs)*

tion chambers and the lightweight hull give peace of mind. Although engines of up to 15 hp can be used, a smaller engine of approximately 5 hp will give more than ample performance.

Stitch-N-Glue kits are offered for ten of the Glen-L designs. It's a practical way to go for beginners. Just about anybody can build a boat using the Glen-L FAST-G system of construction. Many framing and other members that ordinarily require fitting, fairing, and beveling are eliminated. The resulting craft are extremely lightweight, strong, and durable.

Glen-L Marine Designs
9152 Rosecrans
Bellflower, CA 90706
Phone: 213/630-6258
Product literature: catalog, $4

The finished Power Skiff built the FAST-G way is a lightweight frameless craft that's readily cartopped, yet a uniquely shaped boat that's graceful in appearance, stable, and a top performer. *(Glen-L Marine Designs)*

Clark Craft

Clark Craft was the first manufacturer to introduce a prepackaged, do-it-yourself boat kit, and continues to this day to be the world's number-one source of "complete" powerboat and sailboat kits. With sixty-one complete precut boat kits, and precut frame kits for an additional twenty-six designs, the Clark Craft catalog covers everything from kayaks and canoes to a cruising trimarin. Still other designs are offered as plans and patterns only.

Through over forty years of boatbuilding, and as a pioneer in wood/epoxy boatbuilding techniques, Clark Craft has maintained its reputation for innovations with the amateur boatbuilder always in mind. Its recently introduced Epoxy Plus, for a good example, can be used with a thickener as a nonpareil structural adhesive. As a thin coating brushed or rolled on plywood planking, two applications of Epoxy Plus can create a permanent barrier against any type of vapor penetration and add substantially to the strength of a boat hull. If additional strength is needed, an optional layer of glass cloth can be used on exterior surfaces. The Epoxy Plus provides a tough, smooth surface that, when painted over, should last for many years, even on high-speed runabout and hydro hulls.

Clark Craft offers four ways to build a boat: (1) boat kits, (2) frame kits, (3) hull kits, and (4) plans and patterns. The boat kits include all the precut solid mahogany and marine-plywood parts needed to complete the boat of choice, plus fasteners, marine glue, and complete step-by-step building instructions.

Option number two, a custom frame kit, provides the most economical way for the amateur builder to construct a professional-looking hull with preestablished lines, ensuring excellent running characteristics. The frame-

Clark Craft's all-purpose Topper, an ideal boat for cartop carrying, is easy to row, a real performer under outboard power, and, with an optional kit, can be converted to sail at any time. *(Clark Craft)*

kit components, along with the full-size plan or frame-dimension drawings that are provided, enable the builder to assemble the frames, transom, and stem, and establish the basic hull form.

The third option, the hull kit, has become an increasingly popular alternative for the amateur boatbuilder. The concept gives the builder the option to adapt the ideal cabin layout, runabout styling, or fishing accommodations, to a hull of a design and length that best meets individual needs.

We are not interested here in working from plans and patterns alone, except to note that study prints are available for some Clark Craft designs. Study prints, with most costing from $10 to $20, are generally the first sheet of the complete plans set and show the large-scale plan and, in most cases, list the bill of materials. This could provide the would-be boatbuilder with an inexpensive preview of the project and aid in decision making.

We've chosen to feature four Clark Craft designs that are available as complete kits. The first of these is the all-purpose Topper, which can be built in 12- and 14-foot lengths. The 12-footer weighs only 130 pounds and can be lifted by two adults onto a cartop carrier. Topper has all the features of a bigger craft—vee-bottom design, girder construction, full-length spray rails. Easy to row and a real performer with outboard power (up to 20 hp), Topper also converts to a real sailboat. The cost of the Topper 12 boat kit ($499), together with a fastening kit in bronze ($40), an Epoxy Plus covering kit ($122), a basic

(*above*) Designed for high performance, the wide transom and deep tunnel of the 14-foot Tunnel Hull runabout make for softer and smoother riding characteristics in all water conditions—at all speeds. *(Clark Craft)*

(*below*) The Mongoose 20, a 6-meter racer constructed of marine plywood over solid mahogany ribs and frames, using the Epoxy Plus wood-saturation technique, has been clocked at speeds above 70 mph. *(Clark Craft)*

boat hardware kit complete with oar locks and a pair of 6-foot ash oars ($81), and a deck hardware and running lights kit for safe day and night operation ($129), comes to $871. Constructed from the outset as a sailboat, the total kit cost (50-square-foot sail not included) would be $1,141. The sailing option could also be added later.

Designed for high performance, the wide transom and deep tunnel of Clark Craft's Tunnel Hull 14 make for softer and smoother riding characteristics in all water conditions. Construction innovations include built-in flotation tanks in each sponson, additional transom bracing, and double-thickness marine-plywood tunnel planks. The precut boat kit ($795), plus a fastening kit in bronze ($114), an Epoxy Plus covering kit ($185), a deck hardware and lights kit ($129), and a remote-steering kit ($125), packages out at $1,348. Add a power plant with from 25 to 50 hp and you could easily double or triple that price.

If you're looking for pure speed, the Mongoose 20, a 6-meter racer, is designed for sterndrive or outboard propulsion systems of up to 250 hp. Ruggedly constructed Mongooses have been clocked at speeds in excess of 70 mph. The deep-vee racer has a classy ski-boat look. Construction employs the Epoxy Plus wood-saturation system and features 3/8-inch marine-plywood hull coverings over solid mahogany ribs and frames. The cost for the complete Mongoose 20 kit, including bronze fasteners, Epoxy Plus, deck hardware and running lights, comes to $3,090. Figure at least that much again for the propulsion system.

For a do-it-yourself cabin cruiser, there's the Bel Aire 268K Cuddy Cabin, which is one of Clark Craft's most popular designs. The cozy cabin can sleep four and has space for a toilet. The headroom is ample, and side windows open for comfort. The deep-vee-hulled Bel Aire is great for skiing, fishing, or serving as a dive boat. With the cabin to retreat to, it's also a great choice for a family with children who can't take an entire day in the sun

A trailerable 26-foot cruiser, the Bel Aire Cuddy Cabin is one of Clark Craft's most popular designs. It can sleep four, has space for a toilet, and is great for skiing, fishing, or diving. *(Clark Craft)*

afloat. Single or dual inboard, sterndrive, or outboard engines can be installed. The boat kit for the 26-plus-foot Bel Aire Cuddy Cabin costs $5,895. Add another $790 for bronze fasteners and Epoxy Plus.

Clark Craft, Inc.
16 Aqua Lane
Tonawanda, NY 14150
Phone: 716/873-2640
Product literature: catalog, $3

Pygmy Boat Company

The kayaks developed by Pygmy Boat Company combine seven thousand years of Greenland Eskimo sea-kayaking wisdom with the latest in design technology and modern wood/epoxy boat-construction techniques. Less vulnerable than soft-skinned kayaks and lighter in weight than fiberglass craft, Pygmy Boat's Queen Charlotte series of seaworthy touring kayaks are made from precut marine-plywood panels that are literally sewn together and bonded at the seams with epoxy resin and fiberglass tape.

At 38 pounds, the 17-foot Queen Charlotte is one of the lightest sea-touring kayaks on the market. A comparable fiberglass boat would be 15 to 30 percent heavier. Like all Greenland-style boats, the Queen Charlotte has a fine, high bow that parts and rises rapidly through waves in heavy seas. It tracks well and turns easily. The flaring sides give it more stability than most other sea kayaks and the sophisticated hull form and light weight

Extremely seaworthy, Pygmy Boat's kayaks are made from precut and beveled marine-plywood panels that are literally sewn together and bonded at the seams with epoxy resin and fiberglass tape. *(Pygmy Boat Company)*

At 38 pounds, a 17-foot Queen Charlotte touring Kayak is easily transportable. The epoxy-encapsulated Kayak takes about 45 hours to construct and can be handled by a kayaker of any skill level. *(Pygmy Boat Company)*

make it one of the faster sea kayaks. Extremely seaworthy, the craft can be handled by a kayaker of any skill level.

Pygmy kayak kits include everything needed to construct the craft: precut-and-beveled marine-plywood panels, precut cockpit combing, molded fiberglass seat, adjustable aluminum foot braces, epoxy resin and hardener, fiberglass tape, fiberglass cloth, brushes, squeegee, measuring cups, wood flour, rubber gloves, nylon line, and so on. The entire boat, both inside and out, is epoxy-encapsulated. A layer of light fiberglass cloth is used to sheathe the underside of the hull to create an abrasion-resistant surface that allows the boat to be dragged to and from the shoreline. The kayak takes about 45 hours to construct.

Kit price of the 17-foot Queen Charlotte is $550. There's also a 19-foot version of the Queen Charlotte (kit price, $585) for those who want a longer boat. It tracks more strongly, will carry more gear, and has a faster terminal hull speed. At 43 pounds, this is one of the lightest 19-foot singles you're likely to find. The Queen Charlotte XL (kit price, $595), at 17½ feet and with a wider beam and larger cockpit, is designed to accommodate heftier paddlers, and is recommended for anyone weighing more than 225 pounds.

Pygmy Boat has another outstanding design in its Wineglass Dory, which has the wineglass transom and smooth double-ended waterline of a classic rowing boat. Inspiration for the design comes from the "beach skiffs" that were common along the New England coast during the late 1800s. Used year-round in coastal waters, they were made to be launched through the surf and were superb rowers. The 14-foot Wineglass Dory weighs only 90 pounds and, if your only rowing experience has been in heavy, flat-bottomed prams, punts, or dories, you'll likely be delighted at how well a boat such as this rows. But that's not all. The Wineglass Dory, rigged-out with a 63-square-foot sprit sail, centerboard, and tiller, also makes a great sailing dinghy.

The Wineglass Dory is completely computer designed and manufactured. The templates for the sewn-seam boat were cut by a computer-guided laser and are accurate to within four one-thousandths of an inch. While the shape is complex, the exactness of the cuts

(*above*) An all-around skiff with the double-ended waterline of a classic rower, Pygmy Boat's 14-foot Wineglass Dory, which takes its name from the shape of its transom, also makes a great sailing dinghy. *(Pygmy Boat Company)*

(*below*) The Wineglass Dory is computer designed and manufactured. Plank templates for the sewn-seam boat were cut by a computer-guided laser and are accurate to within four one-thousandths of an inch. *(Pygmy Boat Company)*

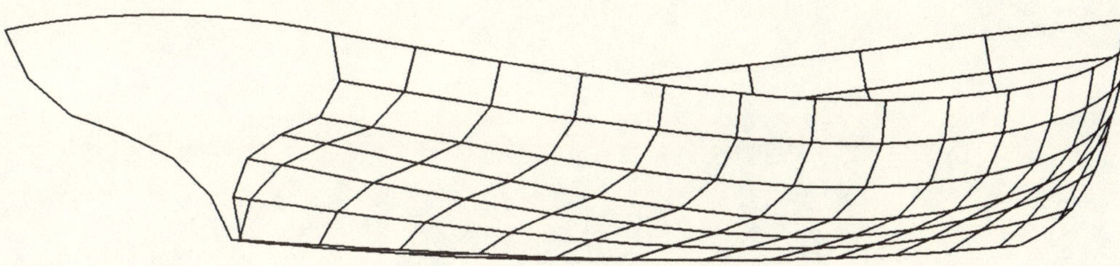

makes the construction process an easy one. The first precut planks are sewn to either side of the bottom panel and temporary transom and center-station frames are sewn into place. The remaining planks are then sewn on. All seams are edge-glued using a dental syringe filled with epoxy. The stitching is then clipped, the temporary frames are removed, and the hull is glassed inside and out. Seats, thwarts, breasthook, knees, transom, and gunnels are then glued in place. The entire boat can be completed in 60 to 90 hours.

The complete kit for the Wineglass Dory costs $890. It can also be purchased as a bare-hull kit ($395), a fiberglass kit ($221.90), and a thwart-and-gunnel kit ($273.10).

Pygmy Boat Company
P.O. Box 1529
Port Townsend, WA 98368
Phone: 206/385-6143
Product literature: free brochure

Paleo Canoe

Factory-made canoes can't hold a candle to the beauty or likely longevity of a handcrafted cedar strip canoe. For looks, there aren't many watercraft of *any* size or shape in the same class as a "stripper." They are built by edge-gluing narrow cedar strips together, and have a long-lasting glass-clear finish that shows off the natural beauty of the wood. Prideful owners of these ribless craft are inclined to list them among their heirlooms, to be handed down to the next generation and the next.

For looks and likely longevity, few watercraft are in the same class as a handcrafted cedar strip canoe with a long-lasting glassed finish that shows off the natural beauty of the wood. *(Paleo Canoe Company)*

For anyone contemplating building a canoe by this basically simple method, whether from a Paleo Canoe kit, some other company's stripper kit, or even from scratch, we strongly recommend first purchasing and reviewing Paleo's 2½-hour step-by-step video, *Building Your Cedar Strip Canoe*. All the know-how is there to guide you from beginning to end. At $39.95, plus $3 for shipping and handling, the video could also be very helpful for any first-timer building a wooden boat calling for a "glassed" hull. The video includes some of the most instructive footage we've seen on applying fiberglass and resin.

Building a cedar strip canoe the Paleo way begins by constructing a "strongback" (basically a long sawhorse) and assembling the wooden form on which the canoe will be built. Paleo supplies the templates for the

Fiberglassing the stripper hull is not without its fascination. As resin is worked into the treated cloth, folds will disappear and the cloth, just as magically, will turn clear as glass. *(Paleo Canoe Company)*

temporary frames, which are cut from plywood purchased locally by the canoe builder. From here, most of the needed materials are included in Paleo's do-it-yourself kits for 14- and 16-foot canoes. With templates, cedar and other woods (including redwood, for a Paleo "signature" hull stripe), fiberglass cloth, epoxy resin, applicators, cane seats, carrying yoke, a copy of the video, hardware, and other items, the kits cost $795 and $895, respectively. This compares with completed Paleo 14- and 16-foot strippers listed for $2,495 and $2,695.

After the form has been readied, the stripping can begin. This consists of edge-gluing long narrow lengths of cedar and stapling them to the wooden frames. After the first few rows, it gets easy. Neatness doesn't count. You can be as sloppy as you want when gluing the strips together. The hull will be scraped down and sanded later and the staples will also be removed. The concern here is to make sure the cedar strips are well-glued and firmly, even though only temporarily, attached to the form.

Once the hull has been glued-up and allowed time to dry, all staples are pulled and the exterior hull surface is scraped down with an ordinary paint scraper. All knot, nail, or other "large" holes are plugged with filler, which is supplied with the kit. After this has dried, the surface is sanded down with progressively finer grits. It is then ready for resin and glassing.

The first application is a sealer coat of resin. This is applied with a paint roller, taking care to roll out all runs. After the sealer has been allowed time to cure, it is sanded lightly and dusted, and the fiberglassing is then begun—with a piece of cloth large enough to cover the bottom of the canoe up to the waterline, and with additional strips to cover the bow and stern chines. Resin is poured on and worked into the fiberglass with a squeegee—and, as if by magic, the resin-impregnated cloth turns transparent, revealing all the beauty of the wood beneath.

A second piece of fiberglass is then applied, with resin, to cover the entire outer surface of the canoe. When this, too, has dried, two

There's a lot of sanding involved in constructing a cedar strip canoe. A belt sander and a pad sander are needed, as well as safety equipment and the accessories needed to capture the dust. *(Paleo Canoe Company)*

more coats of resin are added, sanding lightly between applications.

The hull can now be righted and the form removed. The inside is then scraped, sealed, and fiberglassed with a single layer of cloth. Next, the oak gunnels are added, followed by walnut bow and stern decks. After this, the cane seats and a carrying yoke (or thwart) are installed. Lastly, everything is sanded down lightly with 100 grit, and the entire hull, inside and out, including gunnels and decks, is given two coats of spar urethane with a UV inhibitor.

A 14-foot stripper built from a Paleo Canoe kit will weigh-in at under 50 pounds and have a carrying capacity of 600 pounds. Building one of these canoes *is* labor-intensive and can take anywhere from 100 to 150 hours to complete. But while the end result resembles a fine piece of cabinetry, the "sandwich" construction is very forgiving and ultimately requires more patience than skill.

Paleo Canoe Company
P.O. Box 491
Bedford, MA 01730
Phone: 617/275-3004

CARS

Kit cars give everyone the opportunity to tool around in replicas of machines they could only dream of owning when they were young, originals of which—now that they can afford a few luxuries—have been auctioned out of sight. Most kit-builts today, whether a little Deuce Coupe, a 427 Cobra, or a Jaguar XK120 replica, are so well-conceived that they draw as much attention when out and about as would the real thing. They've indeed come a long way since nearly all replicars were VW Beetle-based and looked it.

Unlike earlier kits, the better replikits today make car building nearly a fun project, with no welding, no home fabricating, and little or no junkyard scrounging involved. Reinforced-fiberglass bodies come with flawless gel-coat finishes in a choice of colors and never need painting. State-of-the-art frames are designed for easy bolting to the running gear of the builder's choice in donor cars. If a needed item is not part of the kit, a source is listed.

Unlike with most kits, the car you build from a kit isn't meant to save you a whole lot of money. If you're just looking for a daily driver, you'd do better shopping the classifieds and the used-car lots. Kit cars, or specialty cars, can end up costing just as much or more than the latest Detroit offerings. Kits succeed because they satisfy a long-time hunger. Slide behind the wheel of a "classic" that you've built yourself, and you'll likely drive around feeling, if not looking, like a million bucks—which could well be the value of the car if only it were an original.

The degree of "authenticity" that can be achieved by the kit builder pretty much depends on how much the builder is prepared to spend. Fiberglass bodies accurately capture the lines of the original, but then it's up to the builder's choice in a drivetrain, as well as in everything from lamps and upholstery to the steering wheel and dash, to capture the feel and performance of the original. Sure, you can drive around in a camouflaged Beetle, but if offered the option of using a VW rear engine or Ford or GM front-engine drivetrain, go with whatever is most appropriate to the original design. There's also no reason why you can't incorporate more recent automotive technology and features—and have a classic with power and handling superior to that of the original.

We strongly recommend that you send for the assembly manual before you make any financially binding decisions in regard to the building of a kit car. Make sure you understand the manual and the car-building procedures in which you'll be involved. If you decide the project is not for you, it's far better to be out the cost of the manual than the cost of the kit.

Some kits call for only the running gear from a donor car. Others call for a whole lot more, right down to the wiring harness and seat belts. So, unless you're looking for a kit to go with a platform you already own, find out beforehand what you'll need from the donor. If you're not all that experienced around cars, you might want to bring in a "lemon detector" to help you with your selection. The body of a potential donor car could be smashed or rusted out, but if your auto diagnostician confirms that the mechanicals are sound, it could be a steal. For donor leads, contact insurance adjusters and check used-car listings in the local newspaper. You might also visit some salvage yards, but only those known to be reputable.

Taking apart the donor car can teach you

a lot that can be applied to the assembling of the kit. There's also the possibility that the donor car may have resalable components. So don't attack the dismantling indiscriminately.

You need room to build a car, usually at least as much space as is provided by a two-car garage. You'll also need a full set of auto mechanics' hand tools. Items such as an engine hoist and jackstands often can be rented, if needed. If you find installation of the electrical wiring or the drivetrain beyond your talents, you might turn to an experienced kit-car builder for help. Many have become either full- or part-time professional kit-car assemblers. They're not only in business to finish cars for others but, with the growing demand for specialty automobiles, to assemble kit cars for resale.

How long does it take to assemble a kit car? Anywhere from 50 to 400 or more hours. It depends largely on the builder's experience and the complexity of the kit. It's a job that shouldn't be rushed, however. The time taken to do things right will only add to the pride and pleasure to be taken in the completed car tomorrow.

The Environmental Protection Agency (EPA) considers the make and model year of a kit car to be the make and model year of the car's underparts—its engine and/or chassis. But state regulations can differ when it comes to registering a kit car. So check the project out with your department of Motor Vehicles early to guard against any surprises.

Bank financing may be possible for your kit purchase. Some kit-car companies also offer competitive bank-rate financing for qualified kit buyers.

Eagle Coach Work

Designed by Sir William Lyons in 1947, the Jaguar XK120 remains one of the most de-

Combining dramatic styling, high performance, luxury, and detail, the Eagle Coach Work XK120G is an updated version of the 1947 Jaguar, one of the all-time favorites of sports-car enthusiasts. *(Eagle Coach Work)*

sired cars ever to have come down the pike. Two feet longer than other British sports cars of its day, the XK120 made the MGs, Triumphs, and Austin-Healeys look like toys. When the big Jag pulled up, you knew the boss had arrived. Some forty-five years ago, Clark Gable bought the first of the XK120s imported into the United States. And today, Eagle Coach Work's updated XK120 replica consistently places among the top three or four when kit-car enthusiasts are asked to rate their preferences from among the more than 150 kit cars currently on the market.

Eagle Coach Work didn't build its first XK120G until 1981, but it has been turning out replica kits for an even earlier Jaguar, the elegant SS100, since 1978. Accurately patterned after Sir William's masterwork of the 1930s, it, too, has been one of the most popular kits ever produced.

The fact that it has concentrated on superb Jaguar replicas is one of the reasons for Eagle Coach Work's success. Another is its policy of selling direct, with no middlemen or commissioned telephone salesmen. This has allowed it to price its kits lower than those offered by the competition, or would-be competition. With over twenty-five years of experience with fiberglass, Eagle Coach Work has also developed an exclusive custom compounding process and delivers its steel-reinforced bodies with a high-gloss gel-coat finish in a choice of eight body colors.

Each SS100G and XK120G is built on a heavy-gauge tubular-steel chassis designed for easy bolting to the running gear of a Ford Pinto, Mustang II, or Mercury Bobcat. Other V8 or V6 engines with either automatic or four-on-the-floor transmission can be used with some modification. Items lifted from the donor car include engine, radiator, front suspension, steering rack, steering column, transmission, drive shaft, rear axle, brake system, wiring harness, heater, and seat belts.

The assembly of both models has been simplified, to eliminate up to fifty steps required to build other SS100 and XK120 replicas. Improved new features include rigid side windows, eliminating the flimsy side curtains of other kits; Carpathian elm burl dash, hand-rubbed and finished with twenty-eight coats of varnish, and added legroom.

There's also a new hard top for the XK120G that most people would swear was a rag top, until they walk up to it and knock on it. It has been designed to simulate the contours of the XK120G's fold-down soft top. All it takes is two minutes for the driver to add or subtract the hard top. With the hard top, rigid side windows, and an exclusive door-latch

With its easily attached soft or hard top and rigid side-window options, the XK120G can give you the best of both worlds—the style of a roadster with the convenience of an all-weather coupe. *(Eagle Coach Work)*

From its Lucas-style lamps to its rear-mounted spare, Eagle Coach Work's SS100 Jaguar replica revives the spirit and elegance of the '30s in one of the most popular replikits ever produced. *(Eagle Coach Work)*

system, the XK120G "roadster" can be locked as securely as any sedan. The hard top is available in all popular colors and comes complete with large rear window, all mounting hardware, electric door openers, padded headliner, and gaskets to ensure a watertight seal.

Price of the deluxe kit plus frame for the Eagle Coach Work XK120G is $8,890. The deluxe kit includes the complete fiberglass body, chrome assemblies, upholstery and trim, and instruments. There's also an option package that includes the elm burl dash, windshield wipers and motor, hand-rubbed wood steering wheel with adapter kit, XK120 hood badge, chrome-plated wire wheel covers, interior mirror, windwings, backup light, windshield glass, license plate bracket and light, and rigid side windows. The price of the option package is $950. For the hard top, add another $1,195.

The deluxe kit and frame for the SS100G costs $8,790. There's an option package similar to the one for the XK120G, also at $950. Eagle Coach Work will sell you a factory-finished SS100G for $25,500, an XK120G for $26,500. You also can order the assembly manual for either kit at $17.50.

Eagle Coach Work, Inc.
760 Northland Avenue
Buffalo, NY 14211
Phone: 716/897-4292
Product literature: brochure, $3

Era Replica Automobiles

In the early 1960s, Carroll Shelby installed a new lightweight 400-hp engine (the 260-cubic-inch V8) from Ford in an aluminum-bodied Ace roadster, built by AC Cars Ltd.

in England, threw in a few tricks of his own, and came up with a racing car that could outrun virtually anything on wheels. The car was the prototype of the now legendary Cobra, and though the early versions were about as aerodynamic as a shoebox, Shelby, in 1965, wrested the much-coveted World Manufacturers Championship away from Ferrari, which for years had dominated GT sports-car racing.

To finance his racing interests, Shelby was turning out street or semicompetition (SC) versions of the Cobra, and it was the highly revamped 427 Cobra (powered by a race-developed Ford V8 engine) that immediately became the ultimate muscle car. The $7,000 "snake," as Cobras came to be called, could go from 0 to 60 in four seconds flat and could top out at close to 165 miles per hour. It was a level of performance that no production car has ever equalled.

By the late 1960s, environmental and other concerns began to apply the brakes to performance and, after the first thousand or so, no more Shelby Cobras* were built. However, for every one of those original Ford-powered Cobras, there are probably fifteen or more replicas. Since the early 1970s, when a Californian named Arntz made the relatively easy-to-clone Cobra one of the first of the specialty cars, its popularity has remained undiminished, and today there are at least twenty manufacturers of Cobra kits or ready-to-roll replicas.

Although not the first to produce a Cobra clone, Era Replica Automobiles, through its quality of engineering, workmanship, and attention to detail, for more than ten years has maintained a solid reputation for producing

*Cobra is now a trademark of Ford Motor Company.

The E.R.A. 427SC may not have been the first Cobra replica but many consider it to be the best. It comes as the most completely engineered and fully detailed kit of its type on the market. *(Era Replica Automobiles)*

While the standard kit is not too shabby, leather buckets, Smiths gauges, competition-type seat belts, and a reproduction of the original Cobra steering wheel are available as upgrade options. *(Era Replica Automobiles)*

the finest Cobra replikits in the industry. The E.R.A. 427SC kit costs $17,900. It's an unusually complete kit and, to control the quality and consistency of finished cars, much of the work that would normally fall to the kit buyer is done at the factory.

The car is a bolt-together, designed just as if it were going to be produced as a finished car on an assembly line. Almost every part is either hinged, latched, or fitted before it leaves the E.R.A. premises. All the holes, caged nuts, receptacles, brackets, and adapters necessary to mount these parts into an in-

Right down to the roll bar, side-pipe exhaust system, and rear nudge bar, E.R.A. goes the extra step to provide snake-bitten enthusiasts with a car as close to the original as possible. *(Era Replica Automobiles)*

tegrated design are provided. The customer, in essence, becomes the assembly line, installing the various components necessary to drive a completed automobile out of the garage.

The major components to be supplied by the kit builder include the Ford engine (big or small block) and transmission, Jaguar XKE final drive/rear suspension unit, Subaru rack and pinion, Triumph Spitfire steering column, exhaust system, and road wheels and tires. Every part necessary to build a complete car is accounted for either in the standard kit or the parts source list.

From the very beginning, the E.R.A. concept was to provide snake-bitten enthusiasts with a car as close to the original as possible, not only in appearance, but in feel, sound, and mechanical layout. The Triumph steering column was chosen because it most closely resembles the column used in the original Cobra; the Jaguar rear drive/suspension unit was chosen for the same reason. As anyone who has built an E.R.A. 427SC repliCobra can attest, this is one wind-in-the-hair car that not only can take you back to another era, but can give you the same exhilarating feeling you would have gotten then.

Era Replica Automobiles
608-612 East Main Street
New Britain, CT 06051
Phone: 203/229-7968
Product literature: 427SC information package, $10

Classic Motor Carriages

The world's largest manufacturer of kit cars is Classic Motor Carriages, of Miami, Florida. Situated on 10 acres and with a 250,000-square-foot production facility, CMC produces kits for everything from Porsche replicas and a "Nostalgia Family" that includes replicas of the 1929 Mercedes Benz SSK (Gazelle), the 1952 MG-TD, and the 427 Cobra, to full-fendered street rods.

Classic's Replicar Assemblies are designed for the person of average mechanical skills, using ordinary tools and the running gear from such less-expensive donor cars as VW Beetles, Chevy Chevettes, and Ford Pintos. Both the Gazelle and the MG-TD assemblies can be ordered for front or rear engine configurations. All fiberglass components come finished in the customer's choice of high-luster gel-coat colors. CMC also gets bonus points for its assembly manuals, which are more detailed than most and filled with helpful drawings.

With its chrome trumpets, oversize lamps, leather hood strap, and wire wheels, the nifty Gazelle combines the design, character, and style of the famous '29 Mercedes-Benz SSK with a modern, dependable chassis and drivetrain. The manufacturer's suggested retail price for the Deluxe VW Gazelle assembly package is $8,495. The package for the Deluxe Ford or Chevrolet Gazelle costs $9,495. All kits include a convertible top with side curtains and fasteners. Instruments from the donor car can be used, but CMC's short list of optional accessories includes old-style gauges, with chrome or brass bezels, priced at $390 and $510 for the set.

The legendary 1952 MG-TD made automobile manufacturers the world over recognize that large numbers of people regarded cars as something more than mere transportation. It continues to be one of the world's most popular and best-remembered cars. Classic's re-creation, for either front or rear engine installations, brings it back in style. Package

(*above*) You'll attract attention and admiration wherever you go in the Gazelle Replicar, which combines the design and style of the '29 Mercedes-Benz SSK with a modern chassis and drivetrain. *(Classic Motor Carriages)*

(*below*) The King of the Road returns in Classic Motor Carriages' re-creation of the legendary MG-TD. The kit for the Classic TD can be ordered for either front or rear engine installations. *(Classic Motor Carriages)*

The Classic rear-engine Speedster captures the perfection of the original Porsche 356A while incorporating many of the advantages of automotive technology developed since its debut. *(Classic Motor Carriages)*

prices for MG-TD assembly kits are the same as for the Gazelle, $8,495 and $9,495.

In the Classic Speedster and the Classic Speedster C, Classic Motor Carriages has incorporated many of the advantages of automotive technology that have been developed since its 1955 debut in its two versions of the sassy Porsche 356A. The Classic Speedster is still so true to Dr. Porsche's 1955 design, though, that about 90 percent of its parts are interchangeable with those of the original. With the Speedster C, CMC has lowered the chassis of the Speedster and added fender flares and racing-type wheels. The replicas are designed for a rear engine only. But then, if you know your automotive history, you know that the original Porsche Speedster was basically a VW on steroids. The simplicity of both models make them among the easiest of all kit cars to assemble. The CMC packages for the Speedster and the Speedster C cost $10,495 and $11,495, respectively.

With the dearth of '30s originals to chop and drop, it should come as no surprise to learn that CMC's blast-from-the-past street rods are among its most popular models today. State-of-the-art tubular-steel frames and one-piece fiberglass bodies with fenders and running boards already attached make them easy to build. Mechanicals are either Ford or Chevy V-8s, and a surprising number of high-tech innovations are included in the assemblies. Package prices for the '33 Classic Vicky

Classic Motor Carriages offers a blast from the past in the '33 Vicky, which features a uniquely designed fiberglass body that includes fenders and running boards already attached. *(Classic Motor Carriages)*

and the '34 Classic Coupe are $12,495. The package for the '34 Classic Cabriolet costs $1,000 more, or $13,495. The kits are remarkably complete, which means parts-chasing will be minimal.

Kits for all of these cars are also offered by Fiberfab International, of Minneapolis. The kits, however, come out of Miami.

Classic Motor Carriages
16650 N.W. 27th Avenue, P.O. Box 10
Miami, FL 33054
Phone: 1-800-CLASSIC
Product literature: free brochure; specify Nostalgia Family, Speedster Family, or Street Rod Family

The custom frame of the '34 Classic Coupe, available with mounts for either small-block Chevy or Ford V-8s, has been engineered for simple bolt-on installation of the drivetrain components. *(Classic Motor Carriages)*

Total Performance

As one of the first and largest street-rod businesses, Total Performance can supply everything from a dropped front axle to a turnkey Deuce Coupe. Founded more than twenty years ago as the outgrowth of the hobbies and dreams of four Connecticut street-rod enthusiasts, Total today is a veritable supermarket of authentic Ford Models T and A replica parts and accessories—as well as kits for some of the most awesome rods and cruisers ever to hit the asphalt.

Specializing in street rods based on Ford cars of the '20s and '30s, Total offers everything from a '23 T Bucket Roadster to four versions of what many consider the ultimate street rod, the classic '32. Any of these soul-satisfying cars can be purchased as complete turn-key vehicles or in various kit stages. The Total Performance catalog includes a 2-page worksheet for each vehicle, and the customer, in effect, assembles an individualized kit, checking off body and chassis packages, parts, and options. A list of labor charges is also included, for those who might prefer to have holes predrilled for a top kit or fender brackets, or to have the upholstery kit or brake lines preinstalled.

The basic kit for the T Bucket Roadster, perennial leader of the TP product line, features a fiberglass '23 T body with molded 14-inch pickup bed and a ready-for-paint jig-welded frame. Basic '23 T kits are priced from $3,500 with a standard front axle, from $3,600 with a performer-style front axle. Then there's a whole laundry list of brake, wheel, upholstery, windshield, fender, head

Basic body-and-jig-welded-frame kits for the '23 T Bucket Roadster, perennial leader of the Total Performance street-rod line, are priced from $3,600 with a performer-style front axle. *(Total Performance)*

(*above*) Components for the nose-down, tail-up Pro Street T, which looks like it's ticketable just standing still, arrive predrilled and jig-welded for hassle-free assembly using basic hand tools. *(Total Performance)*

(*below*) Body and components package for the Pro Street T. *(Total Performance)*

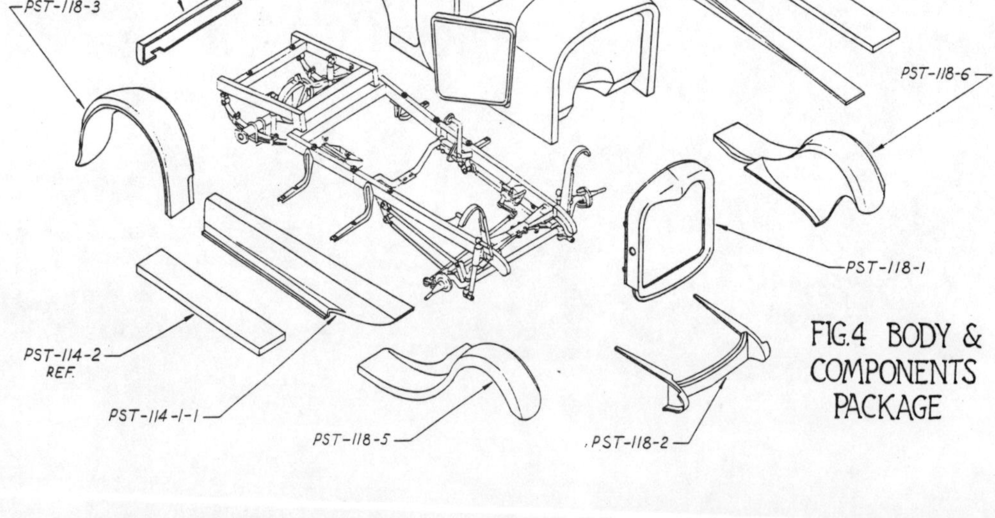

FIG.4 BODY & COMPONENTS PACKAGE

light, fuel tank, and other kits, as well as chrome options. Figure on spending around $9,000, or $11,000 if you want a load of chrome, for a do-it-yourself '23 T Roadster package that will include everything but the engine and transmission. For a turn-key car from Total Performance, prices for the '23 T start at $15,000.

You could also enter the world of hot rodding in "killer" style with the Pro Street T Roadster. Pro Street styling is the hottest of today's trends. Take the finest '27 Ford roadster body available, shave off all the body reveals, tub the rear to accept monster street tires, chop a classic deuce grill shell 4 inches, add Model A running boards—and you've got the Pro Street T cruiser. TP does it all for you in a basic chassis-and-body package kit-priced at $9,500. Whether you build the car up with a mild-mannered four-banger or a 500-inch big block, this is a car that will look like it's doing 100-plus just standing still!

Components for the Pro Street T arrive pre-drilled and jig-welded for hassel-free assembly using basic hand tools. Total outlay, excluding rolling stock and drivetrain, would come to around $15,000, or about half the $30,000 you'd pay for a turn-key version of the car.

Total Performance also offers a full line of street rods based on the Model A, which was first introduced by Henry Ford in 1928. TP's A-line body styles include '28–'29 and '30–'31 roadsters, the '30–'31 coupe, and the '31 two-door phaeton, for either a quaint, resto-rod image or a much more performance-oriented look. Body packages are priced from $3,600 to $5,500, chassis packages from $3,200.

With The Total '32, Total Performance does more than maintain the image of the classic '32 Ford. By offering four versions of

The Highboy Roadster is one of four variations on the classic Deuce Coupe offered by Total Performance. With TP's multitude of options, you could also create your own classic '32 look. *(Total Performance)*

this popular body style—in The Coupe, The Roadster, The Phaeton, and The Highboy Roadster—TP fulfills the wish list of the '32 enthusiast. Total '32 chassis packages are priced from $4,300, body packages from $4,250 (for the '32 Coupe, in either chopped or stock height versions). With TP's multitude of personalized options, you can create your own classic '32 look or build the all-American hot rod without any need for junkyard scrounging or home fabrication. The quintessential street rod is also available from Total Performance as a turn-key car, with prices starting at $40,000.

Total Performance, Inc.
400 South Orchard Street/Route 5
Wallingford, CT 06492
Phone: 1-800-243-6740
Product literature: kits-and-parts catalog, $5

Musical Instruments

Guitar Shop Supply

Stewart-MacDonald's Guitar Shop Supply has been serving the stringed-instrument maker and repairer since 1968. It's the professional's source for everything from plans and body blanks to instrument cases and replacement parts—not just for guitars, but for banjos, mandolins, violins, and dulcimers. It's also the source of a number of exceptional kits for discriminating players and newcomers who want to learn the craft of instrument making and repair from the inside out.

The Guitar Shop's kits are made up with the same high-quality parts supplied to professionals. Its Herringbone Acoustic Guitar Kit, for example, includes bookmatched Indian rosewood back sets and prebent sides, spruce top, reinforced one-piece mahogany neck, profiled ebony fingerboard, and sealed tuners. The most critical steps—neck shaping, fingerboard slotting, soundboard joining, and thickness sanding—are all done for you at the factory.

The cost of this Dreadnought-size, professional steel-string guitar kit is $355. With more and more pop musicians turning off amplifiers and picking up an acoustic guitar, it would certainly seem to be a timely project. An 84-minute video, *Building the Herringbone Acoustic Guitar Kit*, is available for $24.95. For anyone interested in traditional

Guitar Shop Supply's steel-string guitar kit produces a full-size classical instrument with a pure, unadulterated sound. *(Guitar Shop Supply)*

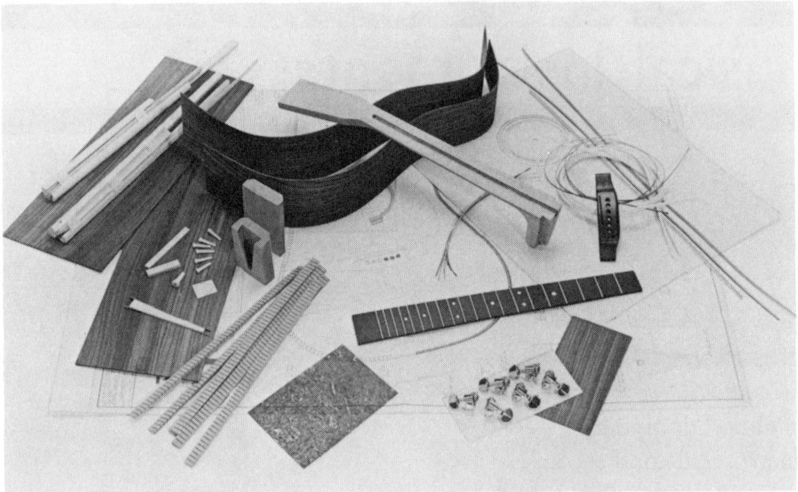

The guitar kit comes with correctly tapered prebent sides; other critical steps also have been done for you at the factory. *(Guitar Shop Supply)*

acoustic guitar construction techniques, it's worthwhile viewing, with or without the kit.

The Guitar Shop also offers kits for "pocket" axes that are fun to build and fun to play. These scaled-down guitars share the same size pickups, bridge, tuners, and neck widths as full-size axes but have smaller Strat, Explorer, and Tele-style bodies and 16½-inch scale lengths (versus the full-size guitar's 25½-inch scale length). Each body is shaped, sanded, routed, and ready for hardware. Spray on a finish, install the parts and wiring, string it up, and you've got a GIANT KILLER. Kit price for the minis is $165. There's also a mini-bass, with a 22½-inch scale. Kit price for the bass is $170.

If your taste is more country than rock, and you recognize such names as Earl Scruggs, Bill

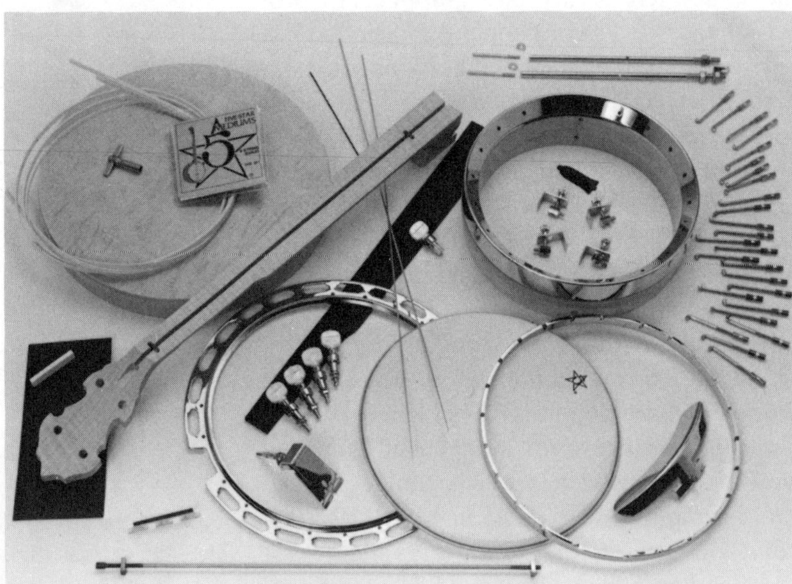

From curly-maple neck and resonator to the Five-Star geared tuners, only the finest banjo parts go into the Bluegrass kit. *(Guitar Shop Supply)*

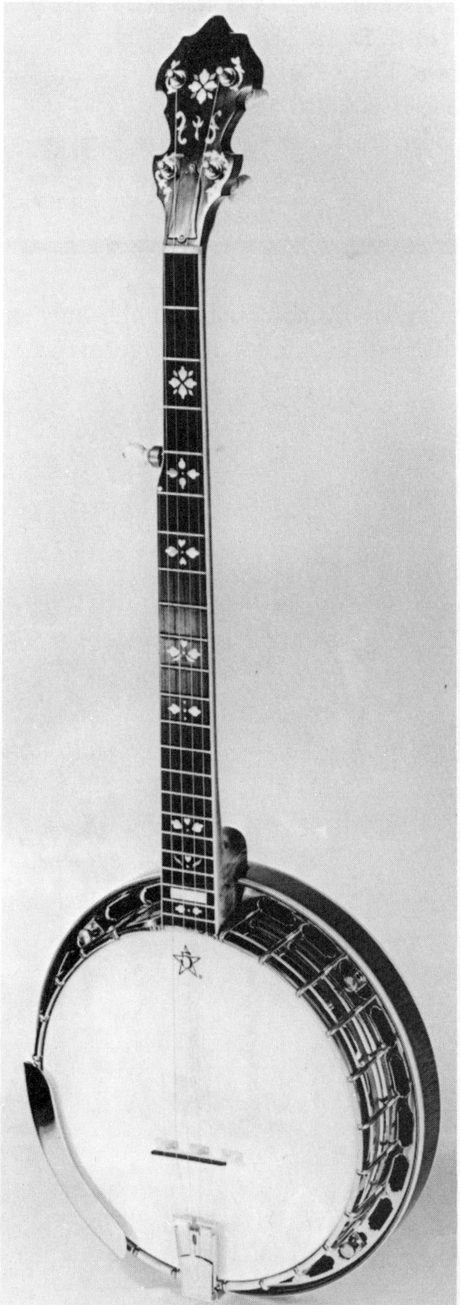

When finished with care, Guitar Shop Supply's Bluegrass Banjo Kit can produce a 5-string instrument that's second to none. *(Guitar Shop Supply)*

Keith, and Sonny Osborne, the Guitar Shop has a Bluegrass Banjo Kit for you that, when finished with care, produces an instrument that's second to none. The Guitar Shop's finest banjo parts go into this kit, which costs $543. Roger Siminoff's photo-illustrated book, *Constructing a 5-String Banjo*, detailing the assembly, finishing, and adjustment of the instrument, is included with each kit.

Another first-rate kit is The Artist Mandolin Kit. Unlike factory-made mandolins, this kit for an F-style mandolin allows you to build a truly custom instrument at an af-

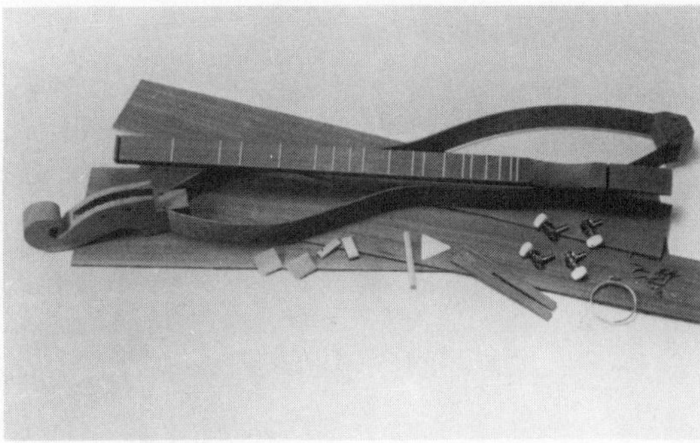

The soundboard and sides of the Hourglass Dulcimer Kit ($108) are bookmatched solid walnut, and the fingerboard is prefretted. *(Guitar Shop Supply)*

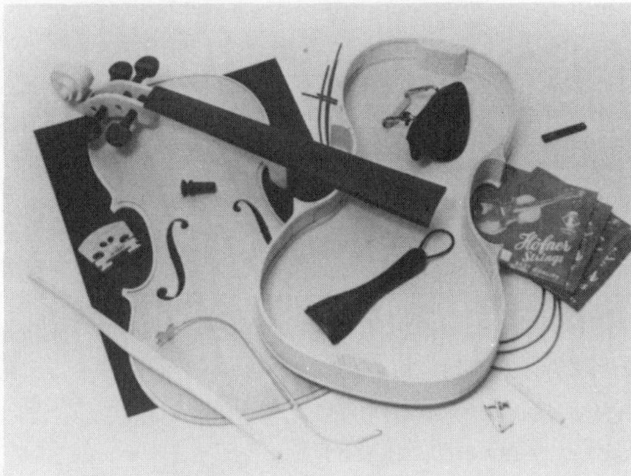

With the spruce top and maple back precarved and sanded, the Guitar Shop's Fiddle Kit ($208) is a comparatively easy project. *(Guitar Shop Supply)*

fordable price, with no compromise in materials. The finest quality precision-machined woods and hardware have been chosen for the kit, which carries a $335 tag.

There's also a fiddle kit ($208), and teardrop ($75) and hourglass ($108) dulcimer kits that offer outstanding value for the money. The dulcimer is one of the easiest and most satisfying instruments to build and play, and either of the dulcimer kits could provide an excellent introduction to instrument making.

Stewart-MacDonald's Guitar Shop Supply
21 North Shafer Street, Box 900
Athens, OH 45701
Phone: 1-800-848-2273
Product literature: free 96-page catalog

Hubbard Harpsichords

It is generally accepted that the only appropriate instruments for the performance of the

The Hubbard Single-Manual Flemish Harpsichord is patterned on an instrument built in Antwerp in 1584, with the 55-note range of the original extended to 58 notes. Precut kit price is $3,465. *(Hubbard Harpsichords)*

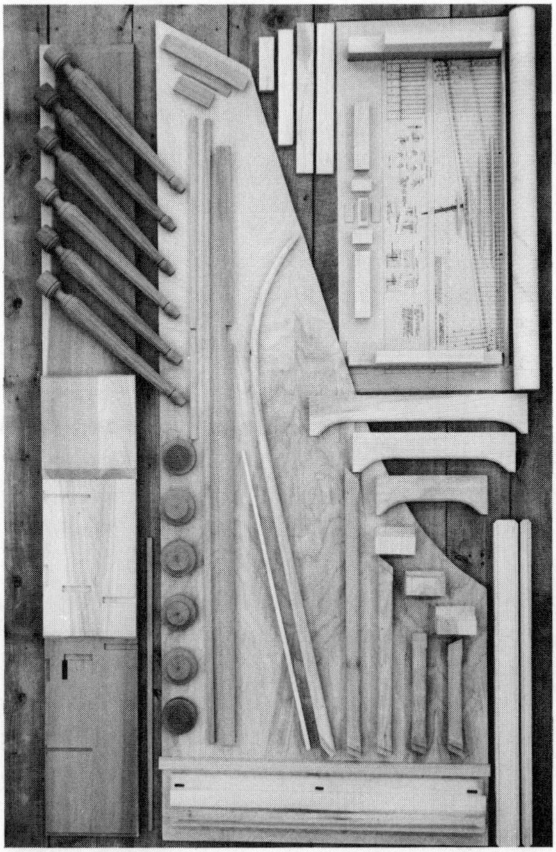

The precut kit for the French Double-Manual Harpsichord costs $4,275 and contains all the items and materials needed to complete the instrument. It takes most builders 4 to 5 months, or about 300 hours of spare-time effort, to finish the project. *(Hubbard Harpsichords)*

music of earlier centuries are replicas of instruments made during the period. Just as baroque music achieves its true color and splendor when played on the harpsichord, so does the keyboard literature of the late eighteenth and early nineteenth centuries acquire its natural brilliance and grace when played on the fortepiano.

Offering kits for harpsichords, spinets, virginals, and fortepianos, Hubbard Harpsichords makes it possible for the ambitious kit maker/musician to acquire a memorable woodworking project and a musical instrument to be enjoyed for a lifetime. If a Hubbard kit is carefully assembled, well-voiced, and handsomely finished, the result equals the product of any professional maker and resembles its early prototype very closely. Hubbard kits have been assembled both by amateurs and professionals and are in use daily by professional musicians in concert halls and teaching studios, and by hundreds of amateurs for their own pleasure.

Designed with the amateur woodworker in mind, Hubbard kit instructions explain in detail procedures for assembly, decoration, and voicing of the instrument, and manage in the process to impart a great deal of information on the art of antique and modern instrument making. There are no short cuts designed to make amateur assembly quick and easy, but with patience and diligence, any amateur woodworker can follow the instructions and produce a thoroughly professional instrument.

All Hubbard kits are offered in several versions, each carried to a further stage of completion, so that the builder may select a starting point depending on his or her resources. With the Single-Manual Flemish Harpsichord, for example, the builder can choose from the precut kit, the assembled case kit (with or without the soundboard installed), and the assembled kit, which leaves only musical assembly and decoration for the builder. The kit prices range from $3,465 to $7,245.

Construction time for the Flemish kit, which is often a first harpsichord kit for the amateur woodworker, is approximately 150–200 hours for the precut kit, exclusive of decoration, which can add 50–100 hours depending on style. The Hubbard Flemish single-manual is patterned on an instrument built in Antwerp in 1584.

Hubbard's spinet is a copy of an English eighteenth-century bentside spinet. It's distinguished from the harpsichord in that the strings and spine run transversely to the player and are not parallel to the key levers. It produces a strong and sustained tone, but one that is darker and less brilliant than the harpsichord. The design of the bentside spinet made it an ideal instrument for musicians with limited floor space. The kit with precut parts costs $2,800.

The fortepiano is a replica of an eighteenth-century Viennese instrument and closely reproduces the tone and action as well as the details of construction of this early piano. Price of the kit with precut parts is $4,920.

Hubbard also offers hinged-seat harpsichord bench kits, with a choice of tapered legs or Louis XVI legs, at $200 and $350, respectively.

Hubbard Harpsichords, Inc.
31 Union Avenue
Sudbury, MA 01776
Phone: 508/443-3417
Product literature: catalog, $3

Musicmaker's Kits

Hammered dulcimers, lap harps, the African thumb piano, and the autochord are but a few of the unusual items to be found in Musicmaker's Keep Book or resource catalog. You may never even have heard of some of these musical instruments, let alone learned how to play them. But then this company goes the extra step and includes in its catalog the best instructional books, tapes, and videos available to help you make good use of your finished handiwork.

As another service to its customers, Musicmaker's rates every project in the Keep Book from 1 to 5, indicating the skill and patience required for successful assembly. Skill Level 1 tells you the project's a breeze. Skill and Patience Levels 5 warn you that the project is intricate and character-building, and requires more than a little perseverance.

If the unassembled kit poses too much of a challenge, many of the kits can be ordered in preassembled and predrilled form. Finished instruments are available as well. If you're a scratch builder, you can even choose to purchase just a blueprint-and-hardware package. The user-friendly catalog tells you what tools are needed and gives finishing recommendations for most kits.

Musicmaker's hand-held Autochord is an easy instrument to learn. Essentially, you just push down the chord buttons and strum. It can be made for right or left hand playing. The kit, with a redwood top, cherry trim, and precut daisy hole, costs $200. *(Musicmaker's Kits)*

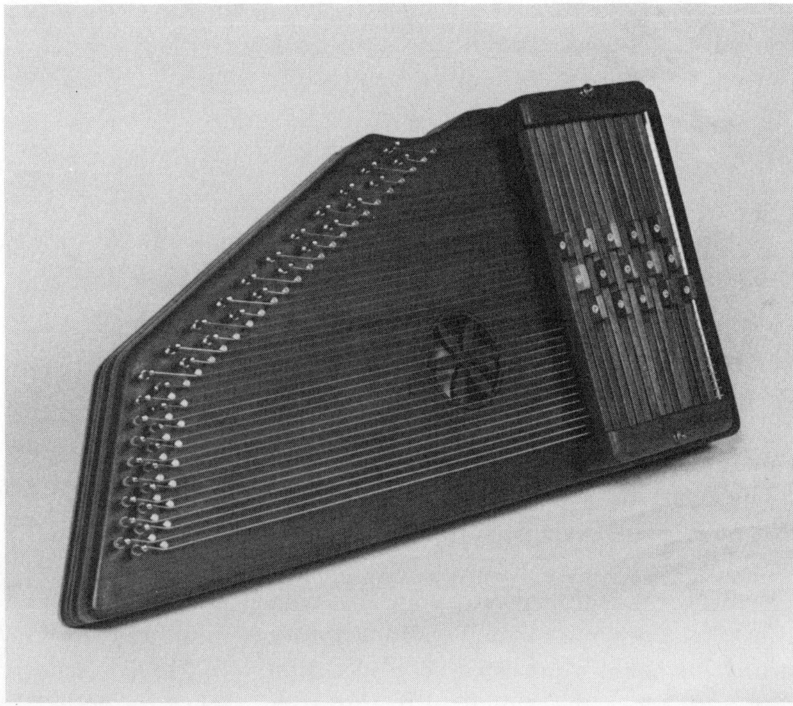

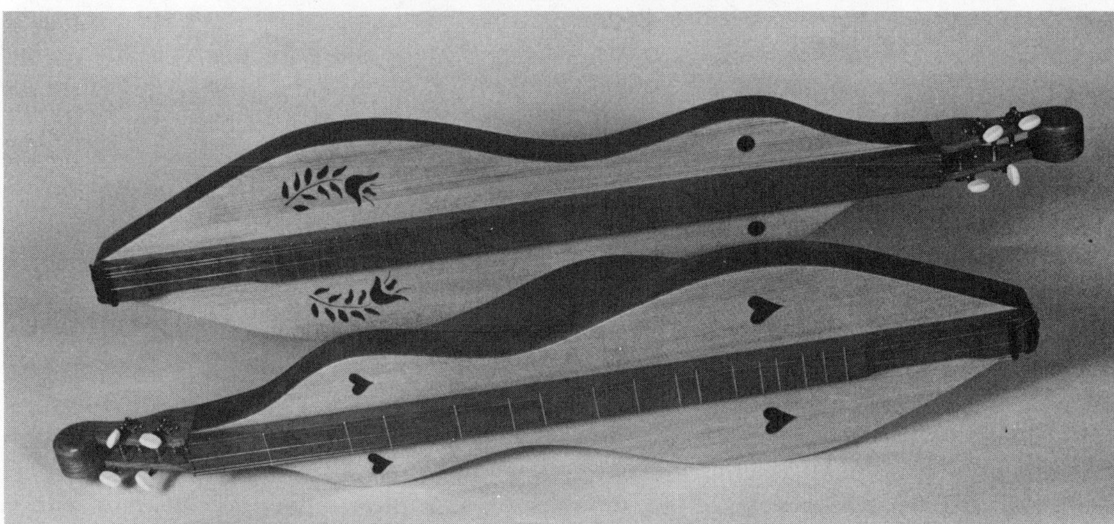

Dulcimers, still played in the southern mountains, are one of the oldest of all folk instruments. You'll love the way your four-string Hourglass Mountain Dulcimer looks and sounds when finished. Musicmaker's kit costs $90 in cherry, $100 in walnut. *(Musicmaker's Kits)*

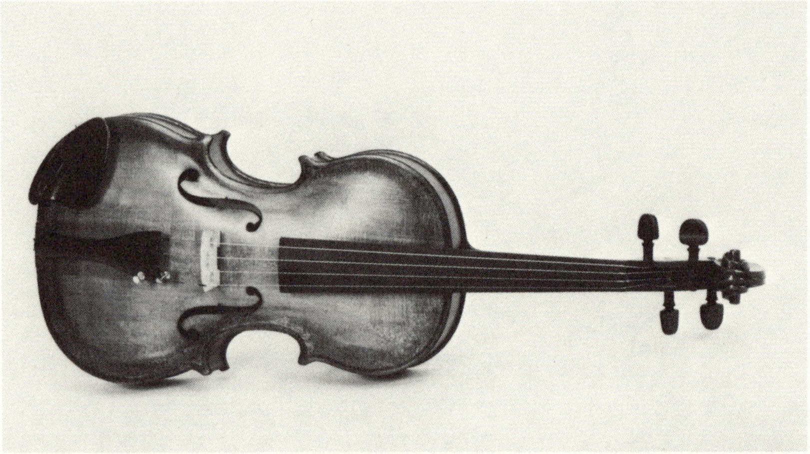

The difficult work has already been done for you in this fine quality German violin kit, making it a surprisingly easy project for the beginner—or a more involved one for the craftsperson who is fussy about soundboard thickness and the final finish. *(Musicmaker's Kits)*

The details on its 5-string mountain banjo provide a good example of what to expect with Musicmaker's. The "voice" is that of Jerry Brown, a dedicated do-it-yourselfer with a love for both music and wood, who has turned his hobbies into this very successful business: "I love this mountain banjo better than the big metal ones . . . It's made of beautiful solid cherry and has a natural goatskin head that has a great sound without chasing people out of the house! [It's] really quite a simple project. All the parts are cut to shape and the rim is already assembled so the prestretched skin head can be dropped into place. You will install the frets into the slots already cut into the fretboard, and you will need to do some drilling and shaping on the tuning head. From there you just sand, finish, and attach the hardware and strings.

"When finishing, I recommend no stain for the wood because the cherry will darken naturally with age. Tung oil would be the simplest and would give a very warm look. Our lacquer finishing kit also works beautifully on this kit, putting more of a sheen on the wood.

"The tools required are a drill, small saw, two C clamps, sandpaper, carpenter's glue, hammer, and a screwdriver."

The mountain banjo is a Skill Level 2, Patience Level 2 project, and the kit costs $139.95. For scratch builders, a blueprint-and-hardware package, including the goatskin head and strings, is available for $65.

Musicmaker's simplest projects include panpipes, in both 10- and 15-tube sizes, and the Little Picker, on which anybody can make music. With the panpipes, which are kit-priced at $24.95 and $39.95, the tubes are carefully tuned before packaging, so you need only lace them to the stave to complete the project. Little Picker (kit price, $24.95) is probably the easiest of all stringed instruments to play. It comes with songsheets and a tuning sheet, with notes positioned under the appropriate strings. You just follow the notes from left to right to pluck the melody. It provides a wonderful introduction for young children, and a great ego boost for the musically hopeless.

Another project the kids can learn with is the 12-note Thumb Piano, which is an African tribal instrument. It's quite easy to make, fun to decorate with paints or inlays, and pleasant to listen to at home, in the car, or around the campfire. The price of this Skill Level 2 kit is $19.95. Surprisingly, Musicmaker's $379 German-made violin kit is also rated as a Skill Level 2 project. But Jerry Brown points out that the majority of construction has already been done for you in this fine-quality kit. It can be an easy project for the beginner, or a more involved one for the craftsperson who is fussy about soundboard thickness and final finish. Jerry said he just assembled the parts as they came from the box and was very pleased with the appearance and sound of the completed instrument. Musicmaker's also offers a high-quality German-made classical guitar kit for $379. It's a skill Level 4 project, however.

Musicmaker's 22-string Irish Lap Harp is only 29 inches tall, 9 inches wide, yet has amazing sound for its size. It's been called the Cadillac of lap harps. The body is walnut, the soundboard birch, and the strings nylon. Kit price is $220. *(Musicmaker's Kits)*

The catalog includes five harp kits, ranging in price from $185, for the 30-inch-tall, 25-string Heartland Lap Harp, to $695 for a 66-inch-tall, 36-string gothic harp. There's also an unusually large selection of dulcimer kits, ranging in price from $59 for the simple Northwoods Mountain Dulcimer, to $275 for the 40-inch-wide, 58-string Performance Hammered Dulcimer. There's even an *electric* mountain dulcimer. This could be Jerry Brown's way of getting the younger generation interested in a timeless traditional instrument! Kit price for the Electric Mountain Dulcimer is $99.95.

Musicmaker's Kits, Inc.
423 South Main
Stillwater, MN 55082
Phone: 1-800-432-KITS
Product literature: catalog, $1

The Early Music Shop

The more than forty different kits produced by The Early Music Shop reflect the growing interest in early music. They are enjoyable and rewarding instruments to build, and each one has been tried and tested, with nothing being left to chance. The Early Music Shop guarantees that, with a minimum of technical ability and skill, you will build a musical instrument capable of the serious execution of the music that was written for it. Even if you're not a serious musician, you may find

The Early Music Shop's small clavichord produces a fine, clear tone capable of a wide dynamic range. Designed to be portable, the instrument weighs approximately 42 pounds, case included. *(The Early Music Shop)*

reward enough in just the building of some of these unusual instruments, many of which can hold their own when set out as a conversation piece.

Home of The Early Music Shop is West Yorkshire, England. Here in America, The Early Music Shop of New England is its representative. Orders for kits as well as the colorful and fascinating 64-page catalog issued by the English company can be paid for in dollars and directed to the New England address. You needn't be concerned with having to deal with British pounds sterling, import duties, or other such complications.

All kits contain every part necessary for their construction, a comprehensive instruction manual and, in most cases, a full-scale drawing with stage-by-stage photographs. You are asked to supply glue and clamps for some kits. The catalog gives an indication of construction time and the degree of difficulty for every kit. For an example, the crumhorn, which is The Early Music Shop's biggest seller, takes 6–10 hours to construct and is rated "fairly easy with a moderate degree of tuning difficulty."

You're not alone if you don't know what a crumhorn is. We didn't either. It's an early

Drum kits include both medieval and renaissance instruments, with goat or calfskin heads and hempen cord and leather thonging used for lacing. Two sticks turned in beech complete each kit. *(The Early Music Shop)*

Within the constraints of lute making, construction of the EMS Renaissance Lute Kit has been made as straightforward and as easy as possible. But it still takes 90–120 hours to complete. *(The Early Music Shop)*

woodwind instrument with a mouthpiece, fingerholes, and a curved main stem. The main stem is pre-bent and a template ensures that drilling the fingerholes is a straightforward process. Even the drills are provided, with concise instructions on how final tuning is to be achieved. The wood throughout is English sycamore. The instruments are available in soprano, alto, tenor, and bass, with either a plastic or a cane reed. Kit prices start at $195 for the soprano-in-c crumhorn.

For something perhaps less esoteric, there are kits for constructing both medieval and renaissance drums, a bagpipe, a hurdy gurdy, and a sixteenth-century lute. During the Renaissance and earlier, the lute occupied a very special place in popular music. No other instrument was referred to more often in Tudor and Elizabethan literature or was illustrated more often in contemporary painting. Construction time for The Early Music Shop's Renaissance 8-course Lute is 90–120 hours. It's "a challenging kit." Traditional materials are used throughout, and the kit costs $600.

The range of instruments represented here, many of which are seldom seen anywhere except in museums, is enough to put even the musically well-educated to the test. There are soprano, alto, tenor, and bass cornamusen; plucked, and soprano, alto, and tenor bowed psalteries; bass and great-bass racketts; a Minster Lovell harp; a cittern; and a soprano shawm. Try a few of those on your musical

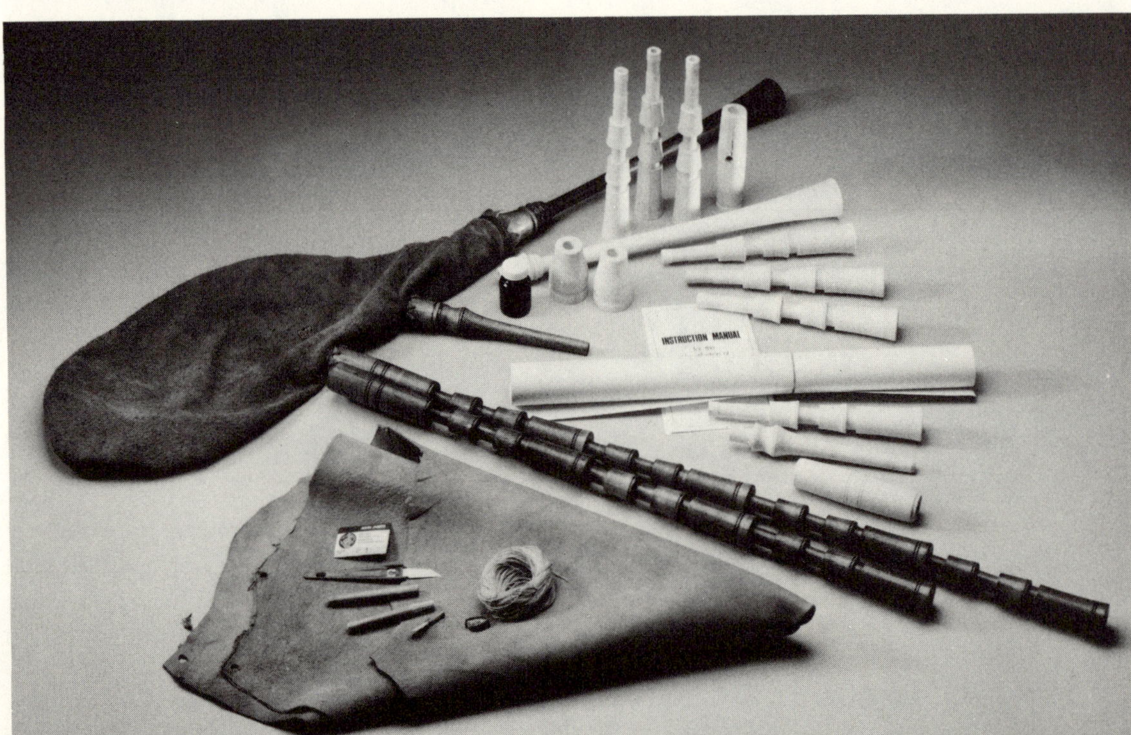

The Bagpipe Kit is easy to build but requires patience to set up. Clear instructions with photographs are given, and the tuning and regulating of the chanter and drones is described in detail. *(The Early Music Shop)*

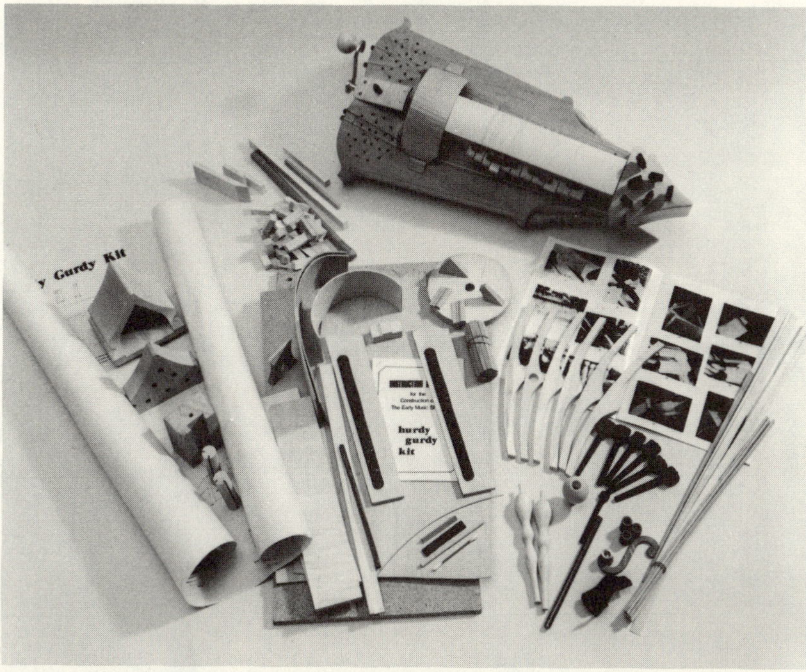

The Early Music Shop's Hurdy Gurdy Kit ($550) is patterned on a thirteenth-century instrument played by turning a crank attached to a resined wheel that scrapes the strings and produces sound. *(The Early Music Shop)*

friends. Kit prices range from $120 for the soprano bowed psaltry to $450 for the cittern and the soprano shawm, which was the forerunner of the modern oboe.

The Early Music Shop's highest prices are reserved for its keyboard instruments, which include a spinet, a clavichord, a cembalino, and a portative organ. The spinet (kit price, $850) is based on an Italian ottavino dated around 1595. The kit builds into a delightful little spinet weighing less than 20 pounds, carrying case included.

The clavichord became popular in the eighteenth century as a less-expensive alternative to the harpsichord. At $1,200, including the carrying case for the finished instrument, The Early Music Shop's small, portable clavichord is still relatively inexpensive. The instrument produces a fine, clear tone capable of a wide dynamic range. A stand kit in mahogany and an attractive paneled lid kit are available as optional extras.

The cembalino kit was produced in response to the requests of Early Music Shop customers for a harpsichord that was easily transportable. The design is based on the compact virginals of the sixteenth century but adapted as a small harpsichord. Even the special clamps required to secure the bentside are provided as part of the kit, and thirty-one color photographs show clearly every stage of construction. Kit price of the cembalino is $1,800. As usual with The Early Music Shop's keyboard kits, the price includes the carrying case.

Rounding out The Early Music Shop's quartet of portable, or portative, keyboard instruments is its Portative Organ, which has a kit price of $2,000. To ensure success even for the first-time kit builder, the bellows for the organ have been specially designed to be simple to make, yet effective for playing. The keys and keybed are predrilled, leaving no chance for error, and the soundboard is fully

Construction time for the Portative Organ is 40–60 hours. The kit has been designed throughout for accuracy and ease of assembly, yet giving a detailed insight into organ principles. *(The Early Music Shop)*

formed with all windways/boring completed. Every screw, pin, dowel, and piece of felt and leather is included in the kit. Construction time for the organ is 40–60 hours. Assembly is straightforward, yet gives a detailed insight into organ principles.

Von Huene Workshop
The Early Music Shop of New England
59-65 Boylston Street
Brookline, MA 02146
Phone: 617/277-8690
Product literature: catalog, $5

The Fischer Collection

If you appreciate both music and fine cabinetry, The Fischer Collection's authentic Old World crank organ offers the do-it-yourselfer a unique combination of aesthetics and sound. The Selberbauen is the world's first mechanical crank organ that comes in kit form, and the musical reproduction from perforated music rolls is of exceptional quality, just as it was in the original European street organs of many years ago. This enchanting project can provide hours of enjoyment during and long after assembly.

The Fischer Collection, which has been building pipe organs for nearly forty years, gets the kits for this hobbyists' dream from the renowned Hofbauer organworks in Germany. All component parts are precisely machined and prefabricated, with all sound-producing parts factory-assembled to assure

This authentic European crank organ in kit form offers an unusual combination of aesthetics and sound. It transports easily if you need a unique gimmick for group entertainment or fundraising. *(The Fischer Collection)*

perfect tonal quality. Cabinet panels are precut to shape, ready for you to assemble and paint or varnish to suit your taste. The kit comes with easy-to-follow assembly instructions and diagrams. No special tools are required.

Small in size (20 inches wide, 27 inches high, and 14 inches deep), but big in sound, the completed crank organ, which is designed to withstand hard use, transports easily and sounds fantastic indoors or out. The musical sounds are generated in the same fashion as the principle employed in much larger theater and church pipe organs.

Price of the Selberbauen kit is $2,660. A dress-up kit, which includes brass corner legs and acorn pipe caps, is offered as an option for $97. A wagon cart, with pneumatic tires and brake, is offered at $498. One music roll comes with the kit, but there's an extensive library of music available, ranging from Viennese melodies and Christmas songs to "New York, New York" and selections from *My Fair Lady*. Additional music rolls, which average six tunes per roll, cost $73.

The Fischer Collection
5295 Old Sterrettania Road
Erie, PA 16506
Phone: 1-800-347-8111

Toys and Other Kits

DOLLHOUSES

You can build a dollhouse as a gift for a child, but a great many of the small-scale houses being assembled from precut kits by both men and women today are for their own enjoyment. Either way, more and more adults are finding that involvement in the world of miniatures gives enormous pleasure.

For some, the dollhouse becomes as much showcase as showpiece, to be furnished with a world of miniaturized collectibles, which

A truly grand Victorian dollhouse, the 40-inch-tall Beacon Hill has become the flagship of the Greenleaf line. No detail has been omitted. There's even a little window in the cellar. *(Greenleaf Products, Inc.)*

The Beacon Hill has seven rooms, and two hallways that are large enough to be furnished as rooms. Three fireplaces and a winding staircase add to the elegance of this outstanding dollhouse. *(Greenleaf Products, Inc.)*

can include everything from a perfectly scaled grandfather clock built from a kit to counted cross-stitch accents and tiny pieces of silverware or pewter bought at flea markets and elsewhere.

Dollhouses, in kit form, come in a wide range of sizes and prices, as well as degrees of difficulty of construction. The source we've singled out here, Greenleaf Products, is the nation's leading manufacturer of die-cut dollhouses. Its products are sold in thousands of shops across the United States; in Canada, the United Kingdom and most European countries; and in places as far away as Australia and Japan. Greenleaf's houses, which include Tudors and Victorians as well as an American series, are noted for their authentic design and reasonable prices. They have a wide appeal to children and to adult collectors and hobbyists.

Dollhouses to Dream Houses, the definitive guide to building and customizing Greenleaf dollhouses, devotes entire chapters to the Beacon Hill and our demonstration project, the Arthur.

A long-time favorite of Greenleaf customers, the 23½-inch-tall Arthur is a wonderful little house appealing to young and old. Its features include silk-screened windows and four ample rooms.

There are kits from one or two other manufacturers that might hold an edge when it comes to the engineering of the kit, but those are mostly hardwood kits priced in the $500 to $600 range. Greenleaf kits in plywood, in the popular one-inch-to-one-foot scale, can cost as little as the $31.92 we paid for the Arthur, the house we've chosen to feature as a demonstration project. It should be noted here that we've seen this kit priced as high as $39.95, and the kit for the Beacon Hill, which has become the flagship of the Greenleaf line, priced from $120.95 to $169.95, so it could pay to shop around for a Greenleaf kit.

Greenleaf houses are constructed of ⅛-inch plywood and, to preserve the sense of scale, various thicknesses of wood are used for trim pieces. Everything is die-cut. As pieces are required in the assembly, they are simply punched out from the wood sheet indicated by the instructions. Assembly is essentially tab and slot, with no tools or fasteners, other than glue, required. With all the gluing, Greenleaf recommends the use of a hot-glue gun, though any wood glue can be used.

Each kit contains all the precut luan plywood and plastic window parts needed to assemble a dollhouse, together with easy-to-read instructions with many illustrations and schematic drawings. Additionally, the kit is packaged with a handsome, full-color illustra-

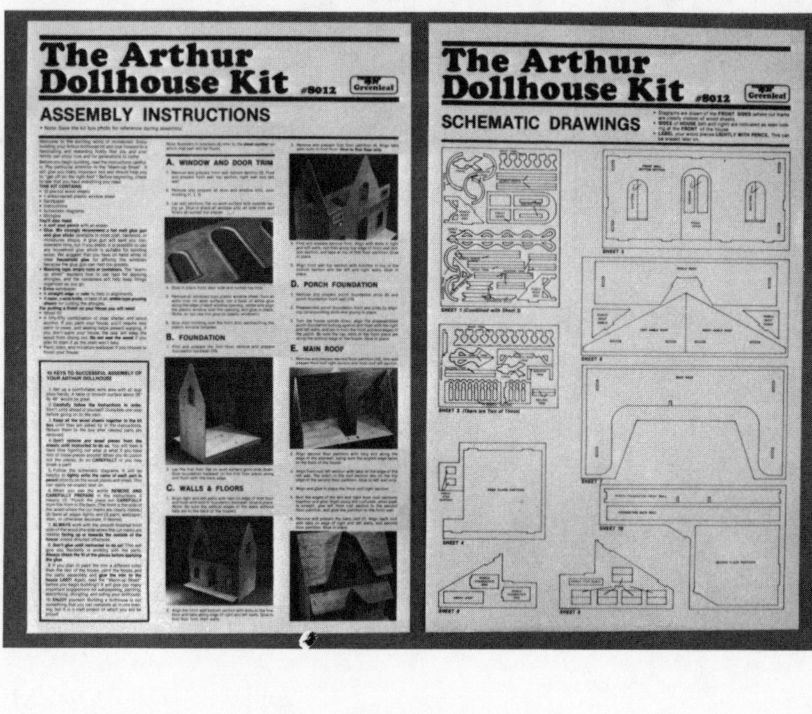

Well-illustrated instructions and schematic drawings printed on 11¼ × 17½-inch sheets leave little chance of error in the construction of the Arthur, if followed in step-by-step order.

Assembly of the Arthur begins with the front and side walls and the gluing in place of all window sills and door and window trim. Silk-screened windows are then added over window openings.

The foundation backwall is installed and the right and left walls are aligned with the first-floor tabs and glued in place. Masking tape can be used to secure joints while the glue dries.

It's important to trial-fit all roof sections to ensure their alignment. If tape is needed to bridge some joints, however, it's not a problem, since it will be concealed by the shingles.

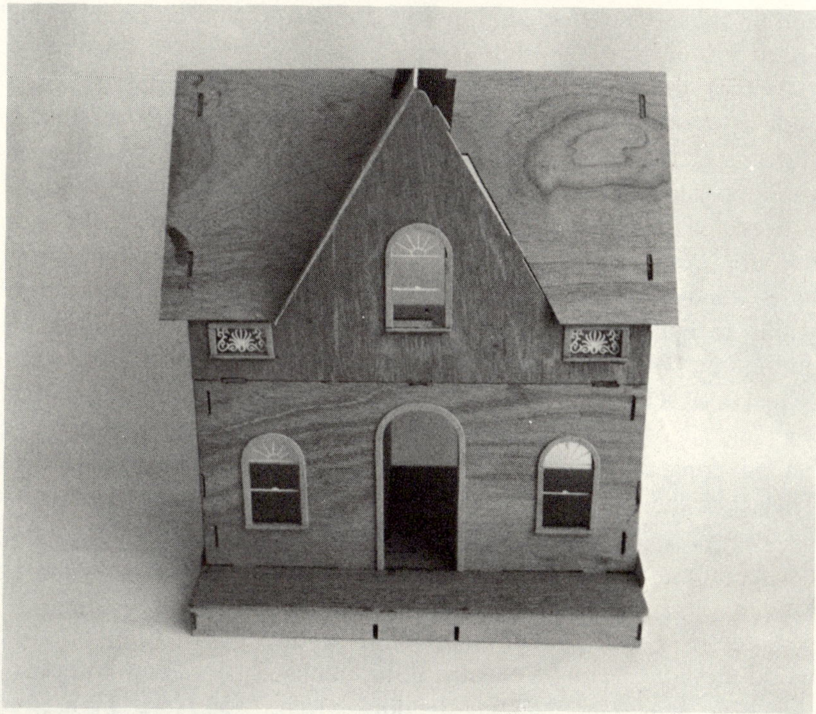

With the front porch installed and shutters and shingles added, all that's left now, should you decide to leave the house raw, is the gingerbread. For others, this may be just the beginning.

tion on the box cover to aid in visualizing the finished product.

Any of the Greenleaf dollhouses can be built as it comes from the box, or it can be customized for an "original," with operable windows and doors, realistic brickwork, and gutters and downspouts. The best guides for this are *Dollhouses to Dream Houses*, volumes I and II, which are the definitive guidebooks to building and decorating Greenleaf dollhouses. The books ($5.25 each) give excellent instructions on every aspect of building and decorating, including electrifying, wallpapering, and painting the dollhouse. They even show how to knock out walls, add on rooms, and build out patios. They also devote entire chapters to individual Greenleaf kits.

Accessories offered by Greenleaf include wood tiles for creating a parquet floor, miniature brass hinges for hanging a door, stucco for ceilings and an authentic look to Tudor exteriors, and a full line of miniature lamps and Elect-a-Lite dollhouse lighting kits. There's even a one-inch-scale die-cut furniture kit ($35) that produces enough pieces (56) to fill a six- or seven-room dollhouse.

These dollhouses include lots of small parts—everything from door and window trim pieces to siding and shingles—so they can take time and do require patience to construct. If you haven't had any experience with dollhouse kits, our recommendation would be to start out with one of the smaller, less-expensive Greenleaf kits, such as the Arthur, or a Coventry or Storybook cottage, before taking on one of the larger Tudors or Victorians. But even as a first-timer, if you use care and don't rush the finishing steps, the result can be a project of which you'll be justly proud!

Greenleaf Products, Inc.
58 North Main Street, P.O. Box 388
Honeoye Falls, NY 14472
Phone: 1-800-847-2545

MINIATURE FURNITURE

Building a dollhouse is only the beginning for many who get hooked on creating their own perfect world in miniature. It's furnishings, complete to the smallest detail, that give dollhouses their real charm and a look of authenticity. A dollhouse *can* be furnished with ready-made scale furniture and miniaturized accessories, but it's much more satisfying to build the furniture pieces yourself. And with quality kits, it's not nearly so difficult as you might think.

The House of Miniatures is the brand most preferred by knowledgeable miniaturists and hobby crafters of our acquaintance. Every piece in its Collector's series is an authentic miniature of an original antique, accurately scaled one inch to one foot. Best of all, each House of Miniatures kit is so well thought out that you will delight at the ease of its completion. The parts are made from fine woods and solid brasses, and the exquisite machining is done in places like Taiwan and Thailand. Otherwise, the kits, which range in price from $3.50 for a Queen Anne candlestand to $11.95 for a Simon Willard tall case clock to $21.95 for a Chippendale breakfront, would be prohibitively expensive.

Every kit includes a layout sheet and hardware checklist showing each part in full scale for easy identification. Step-by-step instruc-

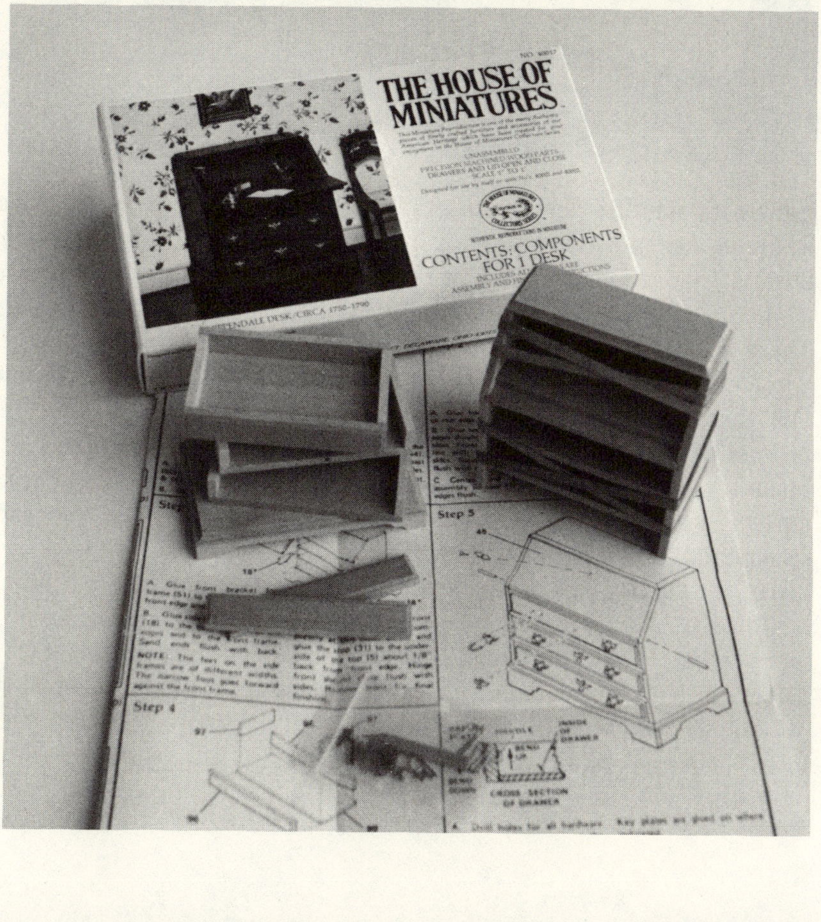

The House of Miniatures exquisitely detailed slant-front Chippendale desk (kit price, $9.95) stands only 3¼ inches tall.

A quarter shows the scale of the Queen Anne tilt-top table (kit price, $4.50). The height of the completed table is 2¼ inches.

tions with detailed views assure that even the beginner will achieve gratifying results. For finishing, we recommend use of the House of Miniatures Professional Finishing Kit ($7.95). There are easier methods, but the end result is what is important—and we haven't found one that gives the quality results that this process does. The finishing kits come in mahogany, colonial, and walnut, and include jars of base stain, glaze stain, clear finish/sealer, and brush cleaner. Several miniature pieces can be finished with one kit.

House of Miniatures kits can be found in many shops where dollhouse kits are sold. There's also a House of Miniatures Collectors Society, through which members receive a different miniature furniture kit each month and are entitled to special membership prices as well as other benefits.

There are more than eighty different true-to-scale House of Miniatures furniture kits available, and whether you are looking for pieces to furnish a dollhouse or to display on a curio shelf or in a shadow box, you are certain to take pride in your accomplishment and will likely treasure each piece that you have made for years to come.

The House of Miniatures, Craftmark, Inc.
147 Lake Street
Delaware, OH 43015
Phone: 614/369-4141
Product literature: catalog, $2

ANIMAL ROCKERS

A toy horse set on rockers and big enough for a child to ride is a traditional gift often given to the first-born of a new family. But you don't have to be a traditionalist to recognize the appeal that Gerry the Giraffe or Daisy the Cow, two of the imaginative designs offered by Animal Rockers, would have for any two-, three-, or four-year-old. The toys not only can provide hours of make-believe fun but can serve to develop balance and coordination.

In addition to Gerry and Daisy, there are Droopy Dog, Ellie the Elephant, Milton the

Precut kits for Gerry the Giraffe and other Animal Rockers include stencils for adding the animal's protective coloration. *(Animal Rockers)*

Daisy and the other rockers are designed to be very stable, but with a 22-inch seat height are recommended for ages two and up. *(Animal Rockers)*

Moose, and Zippy the Zebra. As kits, all hardwood (poplar) parts are precut, shaped, drilled, and sanded. All necessary screws and dowels, as well as manes, tails, and eyes, are provided. The design can either be painted by hand or airbrushed; full-size ready-to-cut stencils are included. Recommendations are given on color selections and the type of (lead-free) paint to use for a durable, long-lasting finish.

All animals measure 22 inches from floor to seat. The fully boarded base, along with footrests and hand-grip pegs, provide for easy mounting and dismounting. Ready-to-assemble-

and-paint kits are priced at $94, plus $14 to cover shipping and handling. Prepainted ready-to-assemble Rockers also are available, at $129.50.

Animal Rockers
P.O. Box 581 Saline,
MI 48176

The New Hampshire Rocking Horse, a careful reproduction of the 1844 original, is offered in kit form by The Bartley Collection (see page 5). In Pennsylvania cherry, with hand-carved head, leather ears and bridle, and seat cushion covered in muslin, the kit price is $319. Seat height is only 13½ inches, making it a safe steed even for the youngest of equestrians. It's one of the most unforgettable presents you could give a special child and sure to become an heirloom. (*The Bartley Collection*)

WHIRLIGIGS

Whirligigs are delightful wind toys and part of our American folk-art heritage. With waving arms, spinning propellers, and sweeping wings, whirligigs take many forms—from birds and airplanes to caricatures of witches and Uncle Sam. The more animated mechanical whirligigs often perform such tasks as cranking cars, chopping wood, or milking cows.

With any of the thirty-six precut kits available from Cherry Tree Toys, you'll have as much fun making and painting a whirligig as you will watching it spin in the wind. Most can be assembled and painted in a couple of hours. Each plan includes an exploded drawing to aid in the assembly of the wind catcher, along with a full-size traceable pattern and color guide to aid in its painting.

No cutting or drilling is required with the precut kits. All that's needed is sandpaper, glue, and a few simple hand tools. Hardwood propeller hubs are solid poplar; dowels are solid maple. The spinning blades are resin-bonded, knot-free Finnish plywood and 100-percent waterproof. Preformed metal parts

You don't need woodworking experience to build a whirligig using a precut kit. Just sand the parts, assemble with glue and the fasteners provided, and it's ready for painting.

A slight breeze is all that's needed to spin the wings of this handsome honker. The precut kit for the 21½-inch goose, or any of eleven other winged wind catchers, costs $19.95.

The tireless woodchopper and other whirligigs can be painted following the pattern and color suggestions provided with each kit or you can add your own decorative details.

are solid brass, bronze, or stainless steel, and will not rust.

Cherry Tree Toys popular wind catchers include twelve different bird forms, ranging from a 15½-inch hummingbird to a 21½-inch-long goose. With these whirligigs, the wings spin and the subject rotates in the breeze. The more whimsical whirligigs include a spinning propeller, rotate freely, and also perform some often-comical, wind-driven action. Kits for the birds, Uncle Sam, a paddling Indian, and other simple spinners cost $19.95. The more complex whirligigs, all of which are 22 inches in length, cost $29.95. Uncut kits are also offered.

Finished whirligigs can be displayed in the yard or garden, or on a display base ($4.95 for the kit) in the home. Cherry Tree Toys also sells plastic resin glue and durable high-gloss, oil-based paints for use on whirligigs and other outdoor toys. The paint, available in eleven colors, costs $3.50 for ¼ pint, $5.50 for ½ pint; the glue costs $1.95 for ¼ pint.

Cherry Tree Toys, Inc.
P.O. Box 369
Belmont, OH 43718
Phone: 1-800-848-4363
Product literature: catalog, $1

DOOR HARPS

Door harps attach to doors and resonate warm, cheerful tones each time the door is opened or closed. Four loosely hung hardwood clapper balls strike the taut musical wires of the door harp at random and create the friendly sounds. Scandinavian in origin, a door harp can add a bit of entrancing hospitality to any home or a business establishment catering to walk-ins.

Cherry Tree Toys, source of the whirligigs on the preceding pages, offers nine different door-harp kits. Measuring from 8½ to 10 inches from top to bottom, these harps have a large, ¾-inch-deep sound box with thin poplar sounding boards on the front and back, for a strong, clear tone. They are easy to assemble and fun to paint. Every kit includes assembly instructions along with an exploded diagram to aid in the assembly and a full-size traceable painting pattern.

No cutting or drilling is required with these kits. You just sand, assemble, and paint. You can either follow the painting pattern provided or add your own decoration, which might be a bit of free-hand folk art or a stenciled design. You could even try wood-burning or adding cutouts. The shape of the door harp itself might inspire the design. There are not only harp shapes, but bell, heart, and wreath shapes.

Cherry Tree Toys offers door-harp kits for all skill levels, from plans and parts-required kits to completely precut kits. Precut kits, which include the predrilled pine blank for the sound box, sounding boards, bridge, predrilled clapper balls, string and fasteners for the clapper balls, music wire, and the eight nickel-plated-steel tuning pegs, cost $17.95. Assembled and unpainted door harps are also available.

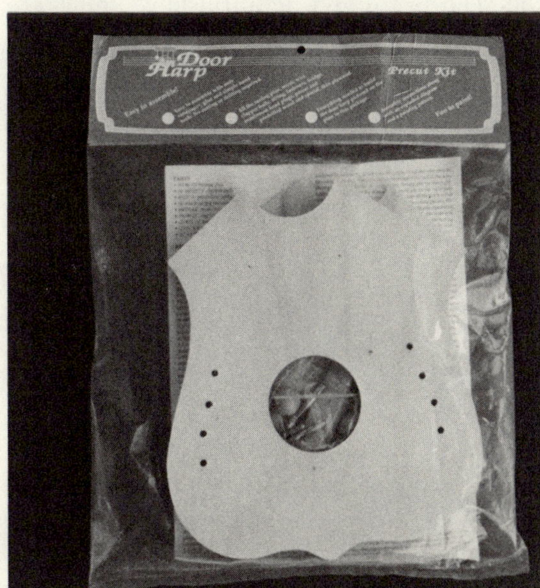

Welcome your guests with the unusual melody of a door harp. The precut kits come in a variety of shapes, with a full-size painting pattern and an exploded drawing to aid in the assembly.

MUSIC BOXES

A music or jewelry box that plays a favorite tune can make a gift that will be cherished for a lifetime. With the music as the message, the gift, with the box handcrafted from a kit, needn't be an expensive one. There are music-box kits, in cherry, and with an 18-note movement included, selling for under $30. If money's no object, however, you could pay as much as $180 for a kit for a rosewood-topped jewelry box, and nearly $400 for a 72-note handcrafted Swiss movement.

Two of the best sources of music-box kits, and the music movements that go into them, are Mason & Sullivan and Klockit, both of which we recognized earlier (see pages 25 and 39) for the excellence of their clock kits. Mason & Sullivan's kits include that rosewood-topped jewelry box. It measures 14¼ x 10½ x 5 inches and comes in cherry, with the rosewood veneer already applied to the case top. Hardware and antitarnishing lining are included in the kit. An appropriate music movement available from Mason & Sullivan for this compartmented case might be a quality 36-note Swiss movement that plays both "Lara's Theme" (from *Dr. Zhivago*) and "Edelweiss." It runs for about eight minutes on a single winding, includes a special start/stop button, and costs $75.

Mason & Sullivan's least expensive music box is a solid-cherry box that measures roughly 5½ x 4 x 3 inches. It comes as a precut kit with an 18-note Sankyo music movement included for $26.95. There are a number of other small music boxes, priced from $39.95

A 50-note Sankyo music movement comes with the larger of the two Mason & Sullivan precut-cherry music boxes shown here, for a package price of $199. The precut kit for the smaller music box, also in cherry, comes with an 18-note movement for $26.95. *(Mason & Sullivan)*

Either of these Klockit jewelry/music boxes would make a gift to be cherished by someone special in your life. The larger case is the 12¼ × 8½ × 2¼-inch Heritage, which comes in walnut and costs $59.95 for the precut kit, music movement not included. *(Klockit)*

to $129, in the Mason & Sullivan catalog. The latter price is for a handsome 5½ x 5½ x 3½-inch box in cherry with an inlaid top and includes a 36-note movement.

Klockit's music boxes include one in black walnut that comes with a lid specially designed to display a favorite 5 x 7-inch photo or print under glass. The precut kit for the 7½ x 5½ x 2½-inch case includes a set-in liner, extension plunger pin (depressed when the box is closed, it stops the music), glass panel, and hardware. It costs $24.95. An 18-note Romance movement by Reuge (Swiss) for this box would cost about $10. Musical selections available include everything from "Love Me Tender" to "Music Box Dancer" and "Brahm's Lullaby."

Other Klockit music boxes, also in walnut, include a little charmer measuring roughly 4½ x 3 x 3 inches. It costs $15.95 and is suitable for one of Klockit's many 18-note Sankyo or Reuge music movements, priced from $5.20 and $9.95, respectively. Klockit's highest-priced kit, at $59.95, would be for its 12¼ x 8½ x 2¼-inch Heritage jewelry/music box. Along with a heap of valuables, this felt-bottom case will accommodate any one of Klockit's 18-, 30-, or 50-note music movements, and has all the qualifications to become a treasured heirloom.

QUILTS

The design of Hearthside's Tulip Garden quilt, a Skill Level 4 kit, dates from the 1840s. The tulips, stems, and leaves are appliquéd down to unbleached muslin, alternating with plain muslin. *(Hearthside Quilts)*

In colonial times, a properly raised young lady was expected to have sewn a dozen quilts before becoming affianced, and would complete a thirteenth, the marriage quilt, when she was engaged. Some of those early quilts, handed down from generation to generation, still survive and are among the most collectible of collectibles today. Even more recently made quilts, from the late nineteenth and early twentieth centuries, often sell for thousands of dollars, not for the comfort they can provide but as art objects.

If you have had to deny yourself one of these hand-me-down pieces of Americana, having never found one that you not only liked but could afford, there *is* an alternative. Hearthside Quilts, of Vermont, offers high-quality quilt kits in many of the classic designs, including Wedding Ring, Log Cabin, Cathedral Window, Schoolhouse Block, and Dresden Plate. There are some fifty different designs, at least one of which, Dutch Tulip, predates the Pilgrims coming to America. You not only can have the look of an authentic piece of history but can put it to its intended, practical use as well.

There's a potential heirloom-in-the-making with every Hearthside quilt kit ordered. The kits are complete, with all pieces for the cotton patchwork top precision-cut to the needed size and shape. Every kit also comes with seamless muslin backing, traditional-weight polyester batting (natural wool batting is an option), and a quilting needle. While Hearthside kits may be machine-sewn and quilted, you should be advised that many quilting purists regard quilts sewn entirely by machine to be a heresy.

All Hearthside kits are graded according to how easy they are to sew. Skill Level 1 kits are the easiest. Even a beginner can sew these designs. Skill Level 2 kits are somewhat more difficult, but still straight stitching. Skill Level 3 kits are somewhat harder due to curved sewing or simple appliqué. Skill Level 4 kits represent the more difficult appliqué designs. All kits are available in infant, twin, double, and queen/king sizes, priced at $49.50, $103.50, $119.50, and $129.50, respectively. Many color choices in both calicos and solids are offered, and if you want extra fabric along with your kit, say for curtains or a dust ruffle, it can be ordered at the same time as the kit.

Hearthside also offers a selection of 22 x 30-inch doll or bassinet quilt kits, at $24.50. These are an excellent size to get children or adults started in quilting. They can make beautiful wall hangings as well.

Hearthside Quilts
P.O. Box 429
Shelburne, VT 05482
Phone: 1-800-451-3533
Product literature: catalog, $3

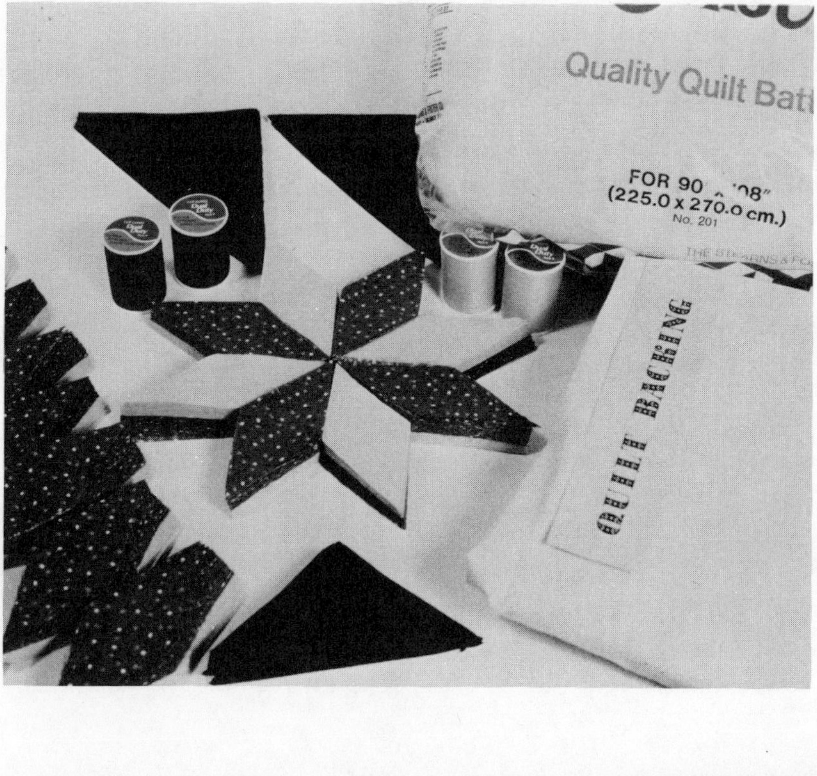

You never need to shop around for materials to complete a Hearthside quilt. The kits come with backing, batting, and precut pieces for the patchwork top. Step-by-step instructions are included. *(Hearthside Quilts)*

SPINNING WHEELS

Craftsman's Corner's fully functional reproduction of a colonial-era spinning wheel can add an unusual touch of authentic Early American charm to a family room, den, or bedroom. Sturdily crafted of solid oak with special pivot points of natural leather, others of metal, you can almost hear the hum of the big wheel as the colonial housewife, or perhaps a daughter, sat near the warmth of the fire patiently spinning yarn which would later be loomed to provide the material for blankets and warm winter clothing.

We won't task you with having to learn the nomenclature of the wheel, other than perhaps to pique your interest by pointing out that the "mother-of-all" and the "maidens" are part of that assembly above the front leg, between the distaff and the big wheel.

The kit includes the precut wood and leather parts, hardware, drive band, and easy-to-follow directions. If the light-colored oak doesn't suit your decor, it can always be stained and finished in a darker color. The diameter of the wheel is 21 inches. The price of the kit is $270.

Craftsman's Corner
4012 N.E. 14th Street, P.O. Box AP
Des Moines, IA 50302-0556
Phone: 1-800-373-1101
Product literature: free catalog

Few reproductions of colonial-era artifacts can add the touch of Early American charm that a working spinning wheel provides. *(Craftsman's Corner, Inc.)*

TRAIL RIDER SADDLE

Tandy Leather has everything needed to make a customized pleasure or roping saddle, even a complete trail rider saddle kit. *(Tandy Leather Company)*

Most youngsters, either in school, scouting, 4-H, or camp, get to try their hand at leathercraft. At some point in your life, you, too, undoubtedly made a wallet or coin purse, a belt, or a case for a pocket comb. The essential materials for those items, whether an uncut hide and lace or a precut kit, very likely came from Tandy Leather Company, which has been in business since 1919 and is the world's largest supplier of leather and leathercraft kits.

We're not really into crafts in this book, but we couldn't pass up calling your attention to the ultimate leathercraft challenge—making your own hand-tooled saddle. Tandy's Trail Rider Saddle Kit ($475) includes precut leather, quick-change buckles, leather-covered stirrups, acrylic-pile-lined Mother Hubbard skirts, rayon girth with attached rings, off billets, hardware, instructions, full-size tooling patterns, and more. The saddle tree, which gives the saddle its form, is the toughest made—one-piece polyethylene (Ralide) with built-in Loop-Lok horn and molded-in ground seat.

If you truly want to ride in style, Tandy also offers gleaming silver accents—horn cap, six conchos, cantle plate, and two fender or corner plates—for a distinctive, parade-worthy saddle. The silver-plated Montana Silversmith Set costs $115.95. There's also an economy set, in German silver, at $59.95.

Tandy also sells everything needed to make a rugged roper saddle as well as hardware packs for truly customized saddles. Hardware packs include all the special buckles, dees, rivets, screws, washers, plates, nails, tacks, thread, latigo lace, conchos, and saddle string needed. Order the style of tree you want and your favorite leather. Add seat padding, stirrups, and cincha, then tool with your own special designs.

There are over three hundred Tandy leather locations nationwide, offering tremendous selections in leathercraft kits and supplies, as well as classes in everything from stamping to saddle making. Check with your local Tandy store for dates and times. Saddle and other leathercraft kits and supplies can also be ordered through Tandy's 100-page color catalog.

Tandy Leather Company
P.O. Box 791
Fort Worth, TX 76101
Phone: 817/551-9779
Product literature: catalog, $2

3-D HOME KIT

Whether you are planning a new home, an addition, or a remodeling project, there's nothing like being able to preview it—to "see" the final results before construction is actually begun. Architects often create a scale model to work on a design and show the client how the finished project will look. There also are computer home-design firms that can do amazing things with a mouse and a monitor. But architectural and computer-design services can bust your budget. A happier solution for many has been an inexpensive kit that gives everything needed to build a detailed, 1/4-inch-scale cardboard model of the home of one's dreams or to confirm that a proposed addition or remodeling project will work.

Design Works' 3-D Home Kit ($29.95) is not only fun to use but will likely bring out creative talents you never knew you had. Depending on the project you have in mind, you might want to start by building a model or partial model of your home. You can then use your model to knock down existing walls, change the roof design, or add a room, deck, new entry, bow or bay window, or skylight. By working with the model, you can explore many different designs for your home—from major additions to new siding materials and patterns.

A 3-D model with a removable roof will let you live with your new home or improvement before actual construction is started. *(Design Works, Inc.)*

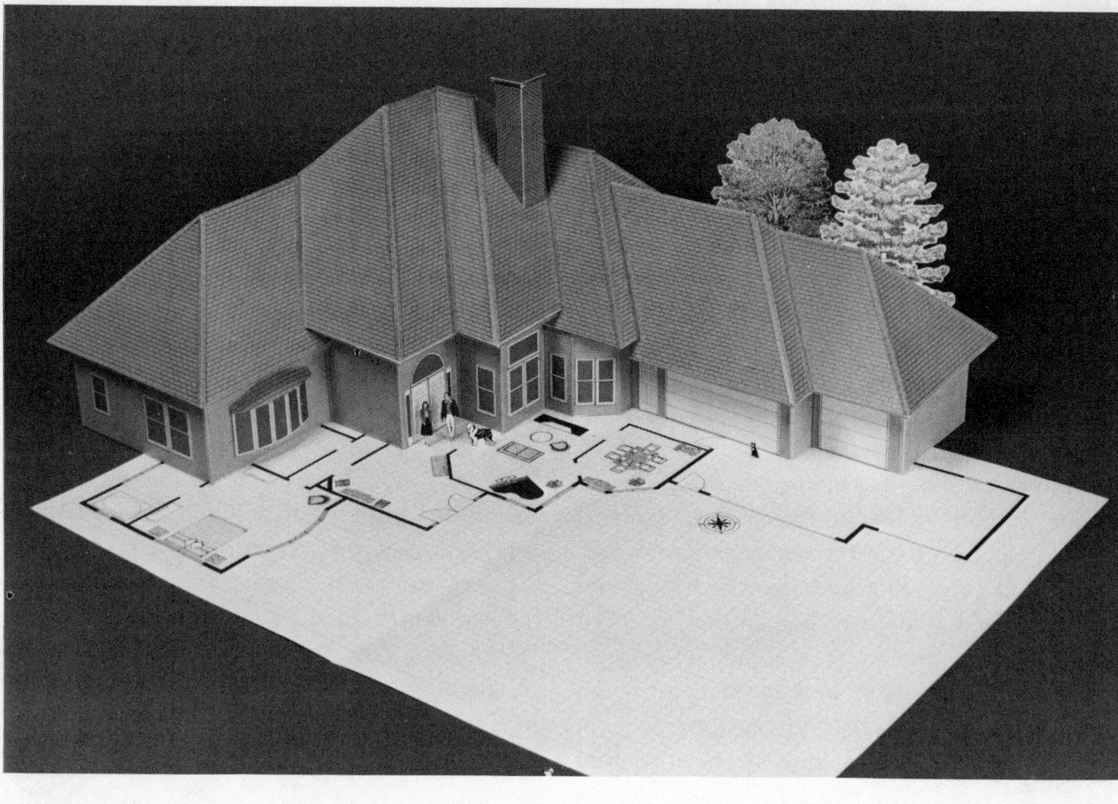

If you're planning to add a bow or bay window or a dormer, or to redo the kitchen, building a model can help you get it right. *(Design Works, Inc.)*

The cardboard building sheets provided include sufficient illustrations of siding, roofing, windows, doors, skylights, decking, brick, stone, interior walls, stairs, kitchen cabinets and appliances, bath fixtures, trees and shrubbery, and even scale humans and a dog and cat, to build and furnish a home of up to 6,200 square feet, or several smaller houses. The kit also includes a floor plan grid and furniture symbols—everything from beds in twin, full, queen, and king sizes, to a pool table and both upright and baby grand pianos—to help organize floor plans, decorating, and interior design.

Notes from a professional designer are included on how to plan and evaluate your design. But bear in mind that *all* plans, whether prepared by yourself or a professional, must be approved by local building authorities before you can proceed with any construction.

Design Works, Inc.
11 Hitching Post Road
Amherst, MA 01002
Phone: 413/549-4763

Index

Note: page numbers in *italics* refer to illustrations.